MXenes: Shaping the Future of Electronic Devices

Mathew

TABLE OF CONTENTS

Chapter 1

Background, Objective, and Structure of this book

MXenes, a rapidly growing family of 2D materials based on transition metal carbides and nitrides, have shown great potentials as multi-functional nanomaterials. Their unique combination of metallic conductivity, hydrophilicity, and highly charged surfaces endow them with excellent electrochemical performance. These characteristics are equally import in electronic and optoelectronic applications of MXenes, henceforth termed MXetronics. MXenes are suitable for solution processing in various polar solvents, allowing the preparation of MXene thin films with controlled transparency, sheet resistance, interlayer spacing, and surface chemistry. The wide selection of various MXene structures, transition metal composition, and surface functional groups enables tunable work function, band gap, and hence electronic and optical properties.

The main objective of this work is to explore and optimize the integration of solution-processed MXene nanosheets into electronic devices. Direct use of aqueous MXene suspension allows simple and cost-effective thin film deposition. Spray-coated MXene thin films with high electrical conductivity, transparency, and flexibility are attractive for electronic applications. The potential of $Ti_3C_2T_x$ MXene as contact material is first studied by integration into n-ZnO and p-SnO oxide transistors and their CMOS inverter, by lift-off patterning method.

However, the conformal deposition of spray coating method results in a side-wall residue of vertical MXene film at the edge of the patterns. This makes a serious issue in multilayer devices, such as high leakage current or device short circuit, limiting their use in the very top layer of the multilayer structure. Hence, dry-etch method was later developed after overcome the lack of strong bonding with the restacked nanosheet thin films and substrate. The hydrophilic surface of MXene thin film becomes a drawback for the treatment with water-based developer solution for photolithography process.

Once the MXene film makes contact with aqueous solutions, the water gets intercalated between the nanosheets by capillary. This reduces the interaction between the nanosheets and make the film vulnerable to the applied strain. By optimizing the strain applied to the heterostructure, the MXene films can be successfully patterned and further integrated into quantum dot transistor. The relatively low work function of MXene compared to vacuum-deposited noble metals such as Au and Pt allows a better electrical contact to n-type channel. In addition to the transistors, the thermoelectric performance of Mo-based MXenes are also discussed.

This book is organized in the following structure:

Chapter 2 provides the basic background knowledge and related literature review of MXenes. Fundamental background of MXene synthesis process, followed by the possible composition of precursor MAX phases and reported MXenes. The surface functional groups of MXenes plays an important role to determine the overall properties of MXenes. This include electronic band structure, work function. The literature review of theoretically predicted diverse properties of MXene family is presented. Based on the restacked nanosheets structure, their interlayer spacing can be engineered for additional functionality.

Chapter 3 demonstrates the utilization of MXene as contact material for n-type ZnO and p-type SnO thin film transistors with their field-effect mobilities of 2.61 and 2.01 cm^2 V^{-1} s^{-1}, respectively. The lift-off patterned $Ti_3C_2T_x$ MXene electrode with work function of 4.60 eV forms good contact for oxide semiconductors with negligible band offset. These individual transistors are also integrated into complementary metal-oxide-semiconductor (CMOS) inverter device that exhibit great gain value of 80 V/V and noise margin of 3.54 V, which is 70.8% of the ideal device at supply voltage of 5 V.

Chapter 4 shows the integration of environmentally-sensitive lead sulfide (PbS) quantum dot semiconductors with metallic MXene electrodes into an electrical double layer transistor (EDLT). The large surface of quantum dots film is effectively gated by ionic gel which also passivate the sensitive quantum dots layer. In order to avoid the side-wall like residue of lift-off approach, a dry-etch method was developed. The successful MXene patterning was the key step to realize high performance EDLT device with large electron saturation mobility of 3.32 cm^2 V^{-1} s^{-1} and current modulation of 1.87×10^4 operating at low driving gate voltage range of 1.25 V, with negligible hysteresis. The relatively low work function of $Ti_3C_2T_x$ MXene (4.4 eV in ultrahigh vacuum) is suitable for n-type transport in iodide-capped PbS CQD films with a LUMO level of ~ 4.14 eV. Moreover, this work demonstrates the first utilization of negative surface charges of MXene for the accumulation of cations at the lower gate bias, achieving a threshold voltage as low as 0.36 V.

Chapter 5 investigates the thermoelectric energy harvesting application of freestanding MXene papers. The samples were prepared by vacuum-assisted filtration of aqueous MXene suspension solution and carefully detached from the filter membrane after drying. The electrical conductivity and Seebeck coefficient of Molybdenum based MXenes (Mo_2CT_x, $Mo_2TiC_2T_x$, $Mo_2Ti_2C_3T_x$) are investigated which had been theoretically predicted as good thermoelectric material. The thermoelectric power reaches 3.09×10^{-4} W m^{-1} K^{-2} at 803 K for the $Mo_2TiC_2T_x$ MXene, which outperform other Mo-based MXenes.

Chapter 6 finalizes the main conclusions of this book, challenges and perspectives for MXene materials for electronic applications are discussed.

Chapter 7 is references for this book.

Chapter 2

Literature Review for the Development of MXenes and Their Electronic Properties

2.1 Introduction

MXene is a large family of two-dimensional (2D) transition metal carbides, nitrides, and carbonitrides that can be derived from 40+ layered $M_{n+1}AX_n$ (MAX) phases or similar precursors by selective removal of the 'A-layers'. The 'M' represents early transition metals, the 'A' stands for A-group elements in the periodic table (mostly group 13 and 14), the 'X' is carbon and/or nitrogen, and n=1-3.[1] MXenes with a general chemical formula of $M_{n+1}X_nT_x$ can be obtained by several chemical routes including hydrofluoric acid (HF),[2] F-containing acidic solutions (typically a mixture of hydrochloric acid and lithium fluoride),[3] or base solutions such as potassium hydroxide for fluorine-free products.[4] The electrochemical route has been recently explored using binary aqueous electrolyte for high-yield and fluorine-free synthesis of $Ti_3C_2T_x$ MXene.[5] The detailed guideline for synthesis and process protocols can be found elsewhere.[6] These etching solutions attack the metallic M-A bonds more rapidly than covalent M-X bonds, hence result in few-atomic-layer-thick MXene nanosheets. While the A-element is etched, the surface of MXene layer becomes chemically active and thermodynamically prefer to create surface functional groups (represented as T_x), such as =O, –OH, and/or minor –F.[7-9]

So far, over 20 different MXenes have been experimentally synthesized and characterized, [1, 10-13] and more are under investigation. The atomic structure of MAX phases and a simplified MXene synthesis process is shown in **Figure 1.1**. MXenes have been under extensive interest since its first discovery in 2011,[2] due to its unusual combination of exceptional properties such as metallic electrical conductivity, excellent volumetric capacitance as supercapacitor, hydrophilicity behavior allowing easy dispersion in aqueous solutions, and the presence of surface termination groups which gives an additional control factor. Most MXenes are predicted to have metallic band structure, while few systems with appropriate surface termination are known to be semiconductor and/or topological insulator.

The diverse electronic properties of functionalized MXenes are summarized in **Figure 1.2**. The large family of MXenes can be first sorted by their crystal structure (single metal type, double metal type – solid solutions or ordered structure, and vacancy ordering type [10-12]) including the number of atomic layers. In addition, the species of transition metal element and surface functional groups are directly related to the electronic properties of MXenes. Although most MXenes are metallic, some semiconducting MXenes are predicted, which are in M_2CT_x formula with oxygen termination and ordered structure. The surface functional groups, interlayer spacing, band structure, and work function of MXenes are discussed in the following sections. MXenes have shown promises in electrochemical energy storage,[1] electromagnetic interference shielding,[14] biomedical applications,[15] and catalysis.[16, 17] The large electrical conductivity of MXenes, which is similar or higher than graphene, will be utilized to novel electrical devices and energy harvesting applications.

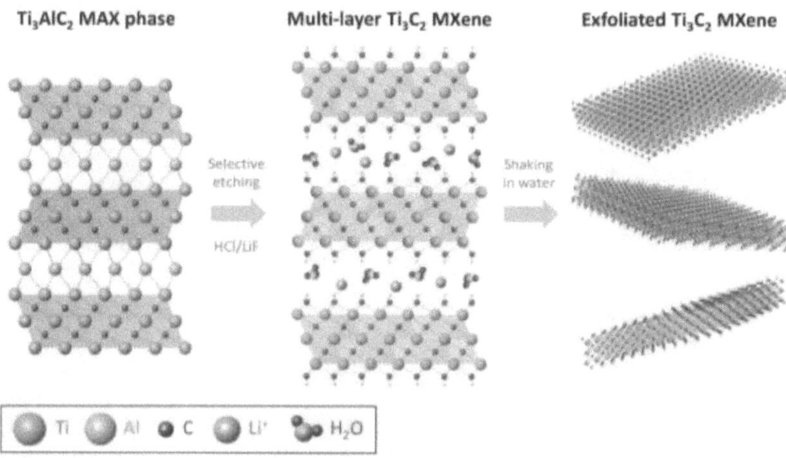

Figure 2.1. Schematic illustration showing the atomic structure of Ti_3AlC_2 MAX phase and corresponding $Ti_3C_2T_x$ MXene after selective chemical etch and exfoliation process.

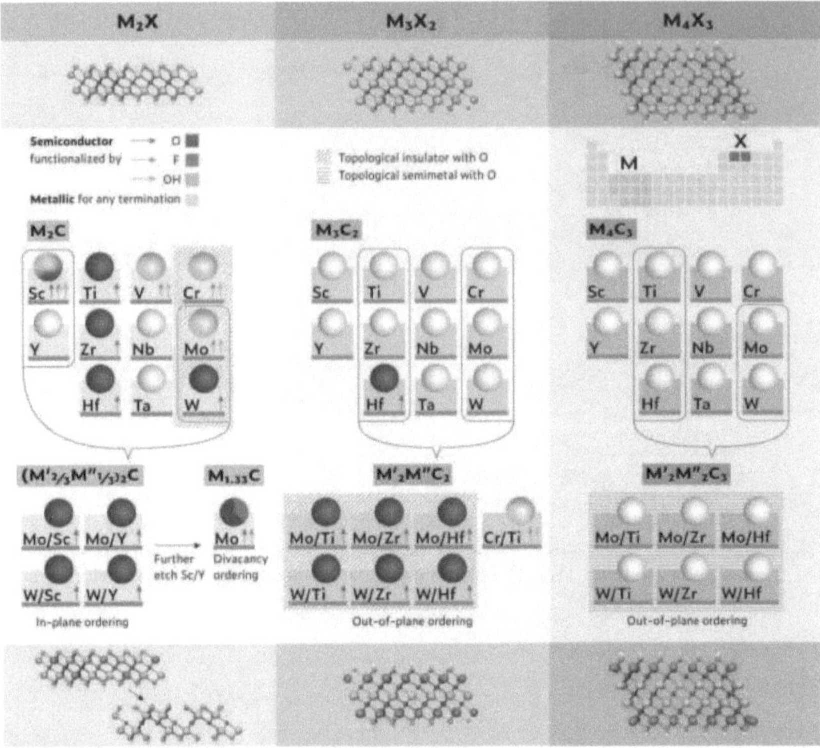

Figure 2.2. Schematic diagram presenting the diverse electronic properties of MXene family. Structure, element of transition metal, and the type of functional groups are considered to sort the electronic properties of MXenes. Semiconductor MXenes are highlighted with colors upon appropriate surface functional groups: blue, violet, and sky-blue colors represent oxygen, fluorine, and hydroxyl group, respectively, while the metallic MXenes with any terminal groups are marked in grey color. The topological insulator/semimetals are shown with slash/horizontal line filled backgrounds, respectively. $Mo_{1.33}C$ is semiconductor with mixed termination of $O_{0.67}F_{0.33}$. Reprinted with permission.[18] Copyright 2019 Elsevier Ltd.

2.2 Synthesis of MXenes

Selective etching methods can be briefly categorized in three ways which use different acidic solutions; (1) HF route, (2) HCl-LiF route, and (3) HF-HCl route. The differences in etch chemistry will be separately discussed in following chapters. After the chemical reaction, the mixture solution contains multilayer MXene which requires further steps to be exfoliated into 2D nanosheets. Typically, the mixture is subject to be washed several time to remove the highly acidic solution and the unwanted reaction products by repeating centrifugation and redispersion into DI water several times. Organic intercalation followed by molecular exchange with water can significantly enhance the yield of MXene nanosheets, however it is difficult to fully remove the remaining organic molecules that usually reduce the electrical conductivity. Ultrasonicaion can induce the delamination of MXene nanosheets, while it also results higher defects density in MXene nanosheets. A schematic illustrion of MXene synthesis process using organic intercalation and ultrasonication is shown in **Figure 2.3**. The different organic intercalants seem to impact the properties of MXene, however it has not been systematically studied yet. A series of selected figures for M_3AC_2 phases, multilayer $M_3C_2T_x$ MXene (after etching), and delaminated 2D $M_3C_2T_x$ MXenes are shown in **Figure 2.4**.

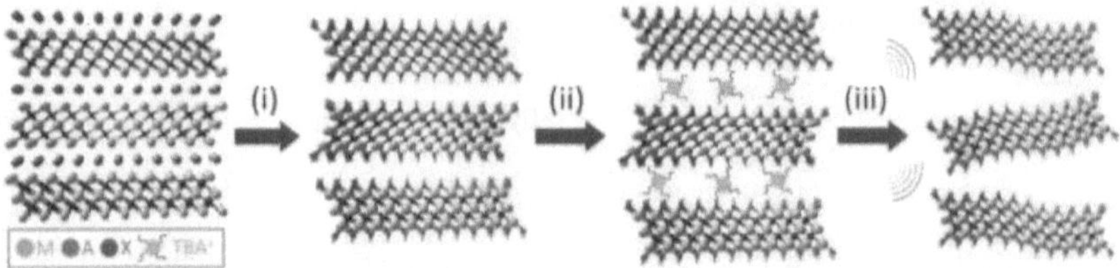

Figure 2.3. Schematic illustration of MXene synthesis process using organic intercalation and ultrasonication. (**i**) Selective etching of A-element from the MAX phases. Red and grey spheres represent surface functional groups. (**ii**) intercalation of organic molecules, tetrabutylammonium cations (TBA$^+$) here, and (**iii**) exfoliation of 2D MXene nanosheets by molecular exchange with water and ultrasonication Reprinted with permission.[19] Copyright 2017 American Chemical Society.

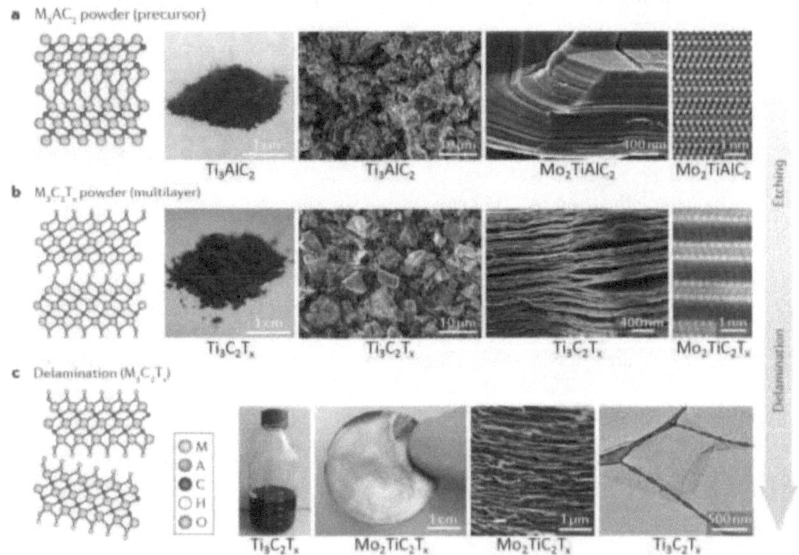

Figure 2.4. Comparison between (**a**) M$_3$AC$_2$ phases, (**b**) multilayer M$_3$C$_2$T$_x$, and (**c**) 2D M$_3$C$_2$T$_x$. From left to right: schematic illustration of atomic structure, digital photography of sample, SEM, cross-section HR-STEM for (a,b), and top-view TEM images for (c). Adopted with permission.[1] Copyright 2017 Springer Nature.

Later an advanced intercalation method using cations is developed, where the small Li^+ cations are to be intercalated into multilayer MXene and result a fine suspension solution upon redispersion into DI water, without ultrasonication. The presence of cations between MXene layer result a hygroscopic nature, and a spontaneous hydration behavior to the relative humidity environment has been reported to be dependent of intercalated cations (Li^+, K^+, Na^+, Rb^+, Mg^{2+}, and Ca^{2+}).[20] The presence of cations between negatively charged MXene also brings co-intercalation of water molecules, which act as lubricants between MXene nanosheets and affect to the c-lattice spacing (c-LP) of resulting MXene samples.

2.2.1 HF route

Hydrofluoric acid (HF) route was first introduced on a selective etching of Al from Ti_3AlC_2, resulting $Ti_3C_2T_x$ MXene in 2011.[2] Due to the highly reactive nature of 50% HF solution, the etching process can be done in relatively short time of 2 hours at room temperature. **Figure 2.5** is SEM images of HF etched multilayer $Ti_3C_2T_x$ MXene showing the typical layered structure with open gaps. The resulting HF-etched MXene is mostly multilayer state without presence of water layer between MXene nanosheets. This can be evidenced by x-ray diffraction patterns. The (002) peak position of Ti_3AlC_2 shifts to slightly lower angle after HF etching by $0.3 - 0.8$ °. The corresponding c-LP of Ti_3AlC_2 is 18.4 Å, and it increases to $19 - 20$ Å after HF etching. The slight increase in c-LP despite of the selective removal of A-layer strongly suggests the formation of surface functional groups, which has been confirmed by density functional theory (DFT) based geometry optimization.[2]

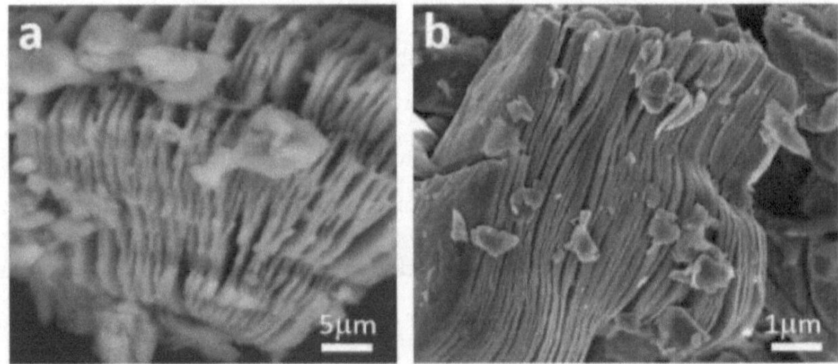

Figure 2.5. SEM images of (**a**) the first reported multilayer $Ti_3C_2T_x$ MXene by HF route. Reproduced with permission.[2] Copyright 2011 WILEY-VCH Verlag GmbH & Co. KGaA, Weinheim. (**b**) Reproduced multilayer $Ti_3C_2T_x$ MXene from our group.

Due to the absence of intercalated water molecule between MXene nanosheets, HF etched multilayer MXene is not readily dispersed into 2D nanosheets. In order to isolate 2D MXene nanosheets from HF etched multilayer MXene, organic intercalation and delamination of MXene has been attempted in 2013.[21] Many organic molecules and solvents (thiophene, ethanol, acetone, tetrahydrofuran, formaldehyde, chloroform, toluene, hexane, N,N-dimethylformamide (DMF) hydrazine monohydrate (HM, $N_2H_4 \cdot H_2O$) dissolved in DMF, dimethyl sulphoxide (DMSO), urea, and long-chain alkylamines) were used for the further treatment of HF etched multilayer MXene, however only HM, HM/DMF, DMSO, and urea were effectively intercalated into multilayer MXene.

Intercalation of HM and HM/DMF at 80 °C for 24 h resulted a further increase in c-LPs from 19.5 Å ($Ti_3C_2T_x$) to 25.4 Å and 26.8 Å, respectively. The latter case shows a synergistic intercalation effect when HM is dissolved in DMF. SEM images of $Ti_3C_2T_x$ before and after HM/DMF are shown in **Figure 2.6**. Upon annealing at 120 °C in a vacuum oven, the c-LP of HM intercalated MXene decreased from 25.4 Å to 20.6 Å, while the c-LP of HM/DMF intercalated MXene decreased from 26.8 Å to 26.0 Å. This is due to the higher boiling point of DMF (153 °C) than that of HM (114 °C). After annealing at 200°C, the c-LP of HM/DMF intercalated MXene decreased to 20.1 Å. The resistance of HM intercalated $Ti_3C_2T_x$ was found to increase an order of magnitude.

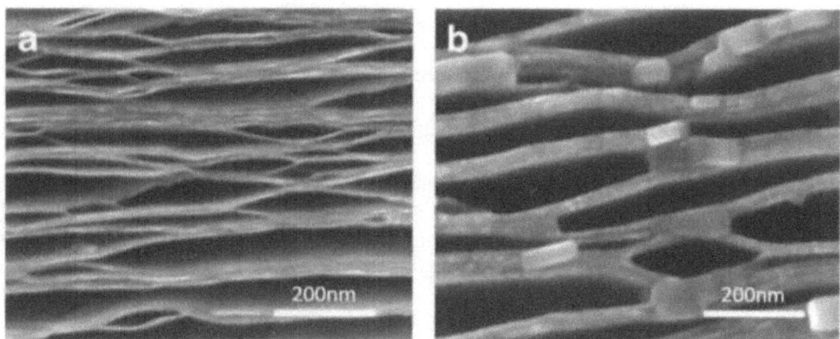

Figure 2.6. SEM images of multilayer $Ti_3C_2T_x$ MXene (**a**) before and (**b**) after the intercalation with HM/DMF at 80 °C for 24 h. The thicker nanolaminate structures are clearly observed in (b). Reproduced with permission.[21] Copyright 2013 Macmillan Publishers Limited.

Intercalation of $Ti_3C_2T_x$ with urea or DMSO resulted in an increase in c-LPs from 19.5 Å to 25.0 Å or 35.0 Å, respectively. When the intercalation reaction of DMSO was extended to 3 weeks, the resulting c-LP shown even larger increase to 44.8 Å. (All XRD tests were performed after vacuum dry at room temperature.) **Figure 2.7 (a)** shows a series XRD patterns of (i) Ti_3AlC_2, (ii) HF etched multilayer $Ti_3C_2T_x$, (iii) DMSO intercalated $Ti_3C_2T_x$, and (iv) freestanding paper using the 2D $Ti_3C_2T_x$ MXene suspension solution. The shift of (002) peaks are clearly observed at each sample state. SEM image of delaminated 2D $Ti_3C_2T_x$ nanosheets on alumina membrane is shown in **Figure 2.7 (b)**. The thin single layer MXene is electron-beam transparent. Size distribution analysis and the Tyndall effect of $Ti_3C_2T_x$ suspension solution is shown in **Figure 2.7 (c)**.

The DMSO intercalated multilayer $Ti_3C_2T_x$ was reported to be highly hygroscopic, and it could be fully delaminated upon sonication in DI water. A freestanding paper consist of 2D $Ti_3C_2T_x$ nanosheets can be prepared by using the resulting suspension solution, which shown four times higher specific capacity for Li-ion battery (410 mAh/g at 1C cycling rate) compared to the HF etched multilayer $Ti_3C_2T_x$. (100 mAh/g at 1C). The capacity of $Ti_3C_2T_x$ paper was higher than that of graphite (280 mAh/g at 1C). This conductive, flexible, and hydrophilic $Ti_3C_2T_x$ paper can host larger cations (such as Na^+, K^+, NH_4^+, Mg^{2+}, and Al^{3+}) by electrochemical intercalation, which exhibit high volumetric capacitances over 300 F/cm^3.[22]

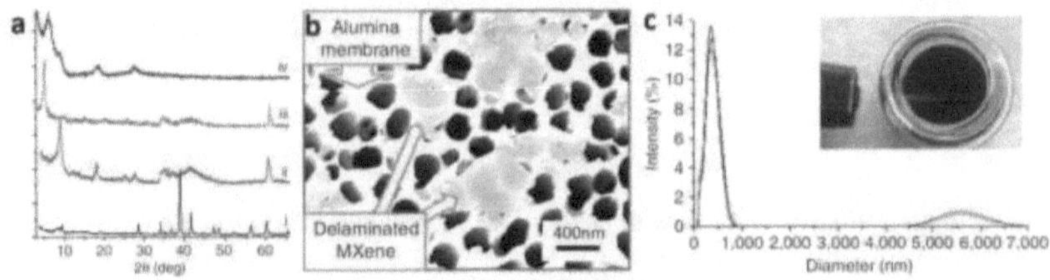

Figure 2.7. (a) XRD patterns of (i) Ti_3AlC_2, (ii) HF etched multilayer $Ti_3C_2T_x$, (iii) DMSO intercalated multilayer $Ti_3C_2T_x$, and (iv) delaminated 2D $Ti_3C_2T_x$ MXene. **(b)** SEM image of delaminated $Ti_3C_2T_x$ nanosheets. **(c)** Particle size distribution of aqueous colloidal $Ti_3C_2T_x$ solution. The inset shows Tyndall scattering effect. Reproduced with permission.[21] Copyright 2013 Macmillan Publishers Limited.

2.2.2 HCl-LiF route

In 2014, a mixture solution of lithium fluoride (LiF) and hydrochloric acid (HCl) was used as a safer alternative etching solution than aqueous HF solution.[3] The presence of Li^+ cation during etching process allows a spontaneous intercalation of Li^+ cations and water molecules into multilayer $Ti_3C_2T_x$. The co-intercalation of water molecule was evidenced by XRD, where the air-dried multilayer state shows c-LP of 27-28 Å without intercalation of organic molecules and hydrated sediments shows even larger c-LP of 40 Å. The XRD pattern of HCl-LiF etched $Ti_3C_2T_x$ are shown in **Figure 2.8 (a)**. The (002) peak is not only increased but it is also sharper compare to HF route. This suggests a better structural order between MXene nanosheets and possible coulombic interaction between negatively charged MXene nanosheets, intercalated cations and dipolar water molecules. More importantly, HCl-LiF etchant is much milder than HF etchant, hence the produced MXene nanosheets have less defects and larger flake size than HF etched samples. Due to the presence of intercalated water, the sediment $Ti_3C_2T_x$ shows 'clay' like behavior and highly conductive nature (~1500 S/cm). By mechanical rolling, an additive-free $Ti_3C_2T_x$ film can be produced which show a large volumetric capacitance up to 900 F/cm^3 (corresponding gravimetric capacitance up to 245 F/g) at a scan rate of 2 mV/s. SEM images of HCl-LiF etched multilayer $Ti_3C_2T_x$ before and after mechanical rolling are shown in **Figure 2.8 (b) and (c)**. The latter case shows shearing which is most probable reason of the loss of non-basal XRD peak.

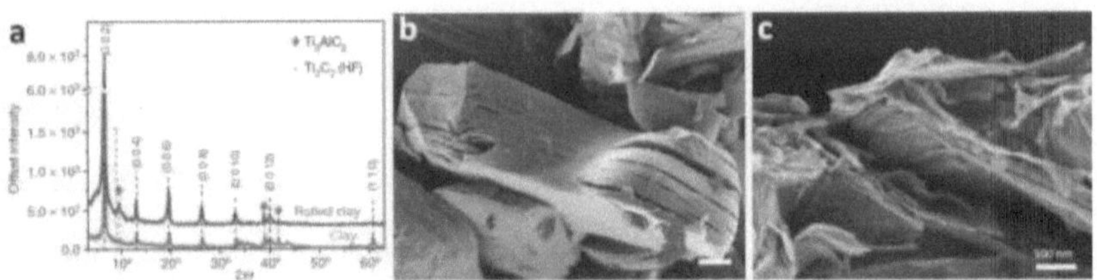

Figure 2.8. (a) XRD patterns of HCl-LiF etched multilayer $Ti_3C_2T_x$ clay and mechanically rolled clay. The (002) peak shifting toward lower angle is clear compare to HF etched $Ti_3C_2T_x$ (green dot line). SEM images of HCl-LiF etched multilayer $Ti_3C_2T_x$ **(b)** before and **(c)** after rolling. Reproduced with permission.[3] Copyright 2014 Macmillan Publishers Limited.

Different fluoride salts (such as NaF, KF, CsF, tetrabutylammonium fluoride, CaF_2) were also tried for HCl etching methods, and all of these salts showed similar etching behavior. Different acids with fluoride salts can also result MXene. The choice of acids and salts for etching step potentially impacts to the surface chemistry and the properties of resulting MXene, hence the etching methods should be further explored.

In a similar method of HCl-LiF, a mixture solution of mild HF (10%) and lithium chloride (LiCl) is reported to show a similar etching behavior.[20] In this case, LiF salt can be formed during etching process, hence the products need to be washed by HCl to remove excessive salt. The presence of LiCl in HF also resulted c-LP of 24.5 Å.

2.2.3 HF-HCl route

In order to increase the yield of 2D MXene nanosheets in suspension solution, an advanced etching method has been developed. By using a mixture solution of HF and HCl, Ti_3AlC_2 MAX can be first etched without using salt. The resulting product can be washed by repeated centrifugation and re-dispersion into DI water, where the absence of intercalated cation and water molecule between $Ti_3C_2T_x$ nanosheets results a minimized material loss during wash process. After washing several times, the pH of solution can reach to ~ 6. The resulting multilayer $Ti_3C_2T_x$ can be re-dispersed into aqueous LiCl solution for Li^+ intercalation. Different salt solutions can be used in this process to differentiate the intercalated cation and surface chemistry of MXene. The cation intercalated multilayer MXene can be washed few times to adjust pH ~ 6 and remove excessive salt ions. High quality $Ti_3C_2T_x$ aqueous suspension solution can be obtained with a concentration of ~ 3 mg/ml, and the vacuum filtrated $Ti_3C_2T_x$ freestanding paper shows a high electrical conductivity of ~ 10,000 S/cm.

2.3 Composition of MAX phases and reported MXenes

Up to date, more than 60 different composition of MAX phases have been reported. **Figure 2.9** is a periodic table which show the elements groups consisting MAX families. The detailed list of MAX phases can be found in literature.[1, 23] Although the A-layer element should be removed in order to obtain MXenes, the A-layer element is still important as it affects to the etch chemistry hence the properties of resulting MXenes. Most MXene studies deal with selective etching of Al layers at current stage, but few of them involves with Ga layer. MXenes can also be obtained using alternative MAX-like phases, for example, $Hf_3Al_4C_6$ (alternative stack of Hf_3C_2 and Al_4C_4 layers) has been used as precursor material to synthesize $Hf_3C_2T_x$ MXene after silicon alloying into Al_4C_4 layers followed by chemical etching.[13]

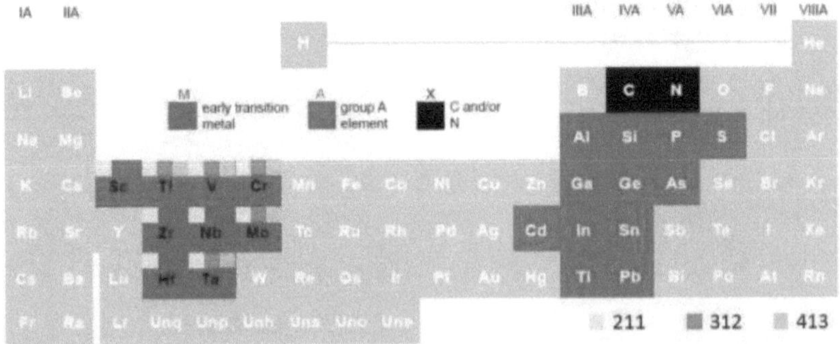

Figure 2.9 Periodic table showing elements which consist MAX phases. Reproduced from with permission.[23] Copyright 2001 Sigma Xi, The Scientific Research Society.

In addition, these MAX phases should not be limited to the mono-elements of each M, A, and X subgroups. Solid solution of at least two different M elements can create same crystal structure. Many solid solution MAX phases are reported such as $(Ti_{0.5}V_{0.5})_3AlC_2$, $(Nb_{0.5}V_{0.5})_2AlC$, $(Nb_{0.5}V_{0.5})_4AlC_3$, $(Nb_{0.8}Zr_{0.2})_2AlC$, etc.[24] A Few systems with two certain M elements in MAX phases can be categorized as ordered double transition metal MAX phase, where one transition metal occupies the outer layers and another locate at the central layer. (For example, Mo_2TiAlC_2 and $Mo_2Ti_2AlC_3$. Mo atoms preferred to locate at outer M layer and Ti atoms tend to occupy the central M layer.[25]) These ordered structure has been predicted to remain in the state of MXenes as well, and this discovery expended the possible combination of MXene family.[26]

Up to date, 21 different MXenes has been successfully synthesized.[1, 10, 13] **(Figure 2.10)** One of the most recently reported MXene is non-stoichiometric $Mo_{1.33}CT_x$, which has ordered vacancies in the lattice of Mo_2CT_x.[10] In this case, quaternary $(Mo_{2/3}Sc_{1/3})_2AlC$ MAX phases with in-plane ordering was first prepared, and both Sc and Al elements were selectively removed by HF, followed by TBAOH intercalation assisted delamination of $Mo_{1.33}CT_x$ MXene nanosheets. A schematic illustration of this approach using sacrificial M layer in shown in **Figure 2.11** along with STEM images of the precursor and resulting MXene. As this recently developed approach can expand the MXene family, many different composition of MXenes are expected to be realized in near future. Such engineered vacancy in MXenes is potentially useful for energy storage and/or hydrogen storage applications.

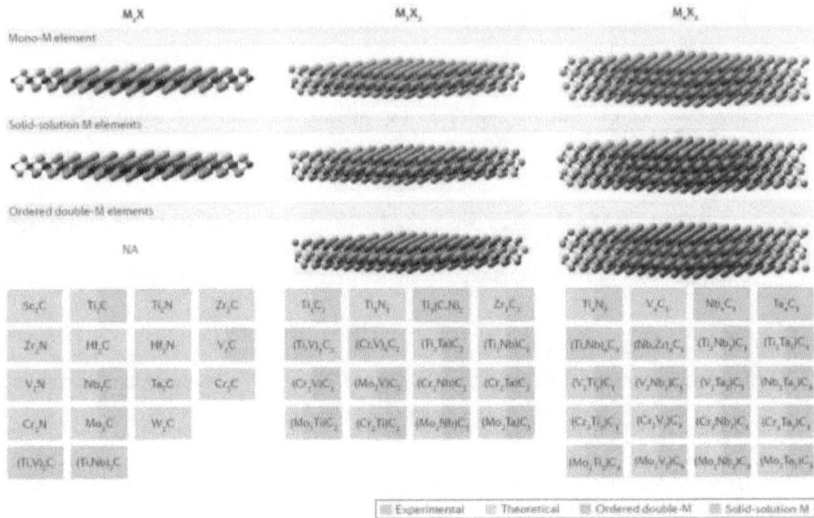

Figure 2.10. Different types of MXenes; M_2X, M_3X_2, and M_4X_3. Schematic illustration of atomic structure for the cases of mono-M element, solid-solution M elements, and ordered double M elements MXenes. Theoretically predicted MXenes to be exist and experimentally synthesized MXenes are marked by grey and blue color, respectively. Reproduced with permission.[1] Copyright 2017 Springer Nature. Recently reported $Hf_3C_2T_x$ and non-stoichiometric $Mo_{1.33}CT_x$ MXenes are not listed on this diagram.[10, 13]

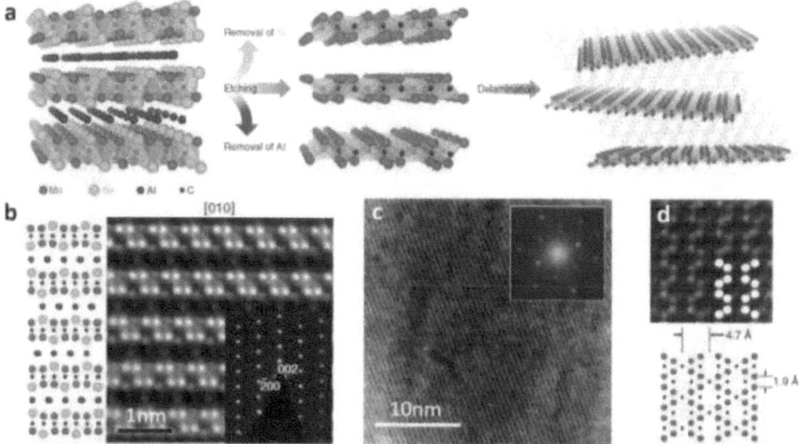

Figure 2.11 (a) Schematic illustration of $Mo_{1.33}CT_x$ MXene synthesis process. **(b)** In-plane chemical ordering of $(Mo_{2/3}Sc_{1/3})_2AlC$ MAX phase with STEM image along the [010] zone axis. The inset is selected area electron diffraction (SAED) pattern. **(c)** Top view of HAADF-STEM of single $Mo_{1.33}CT_x$ MXene in high magnification with Fast Fourier Transform (FFT) image. **(d)** Atomically resolved zig-zag structure with schematic atomic structure which was theoretically simulated from parent MAX phase. Reproduced with permission.[10] Copyright 2017, Springer Nature.

2.4 Basic Properties of MXenes

2.4.1 Surface Functional Groups

The presence of the surface termination groups renders the surfaces of MXenes hydrophilic (water contact angle of 21.5 – 35°). [3, 27, 28] Furthermore, most MXenes are highly conducting (~10^4 S cm^{-1} for the most well-studied $Ti_3C_2T_x$).[29, 30] This unique combination of hydrophilicity and metallic conductivity is one of the advantages over graphene and its derivatives. In addition, the large surface area inherited from their 2D structure makes MXenes promising in electrochemical energy storage applications. The surface of functionalized MXenes is negatively charged due to the large electronegativity of O or F atoms in terminal groups, thus restacked MXene can allow reversible electrochemical intercalation of cations including Na$^+$,[31, 32] as well as multi-valence Mg^{2+} and Al^{3+} by strong Coulombic interaction between MXene and cations.[22] The surface groups also play an important role in pseudocapacitive redox reaction where oxygen groups can be converted to hydroxyl groups while accommodating both proton and electron in H_2SO_4 electrolyte.[33] The stable dispersion behavior of MXenes in polar solvents such as water, ethanol, acetone, acetonitrile, N,N-dimethylformamide (DMF), dimethyl sulfoxide (DMSO), N-methyl-2-pyrrolidone (NMP), and propylene carbonate (PC) can allow facile integration with a variety of solvent-sensitive materials.[34] $Ti_3C_2T_x$ MXene electrodes in supercapacitors with macroporous, hydrogel, and vertically aligned liquid crystal structures have been reported showing excellent electrochemical performance including ultrafast operating rate at 10 V s^{-1},[35] ultrahigh volumetric capacitance of 1500 F cm^{-3},[35] and thickness independent capacitance,[36] respectively.

The surface functional groups of MXenes affect not only their electrochemical performance, but also electronic properties (including band structure, work function), and optical properties. Although the method to precisely control the surface functional group species is yet to be discovered, those surface functional groups are highly dependent on the synthesis route and post-synthesis treatments. ^1H and ^{19}F nuclear magnetic resonance (NMR) spectroscopy study has successfully quantified $Ti_3C_2T_x$ MXenes prepared by HF route and HCl-LiF route.[37] **Figure 2.12 (a) and (b)** show the different morphology of multilayer $Ti_3C_2T_x$ MXene prepared by volatile HF route and milder HCl-LiF route, respectively. The surface functional groups of $Ti_3C_2T_x$ MXene has been experimentally observed as shown in **Figure 2.12 (c) and (d)**, with the schematic illustration in **Figure 2.12 (e)**. The larger-sized MXene nanosheet ~10 μm can be obtained by the milder method **(Figure 2.12 (f))**.

Treatment with tetrabutylammonium hydroxide (TBAOH) has been reported to reduce the fluorine content and increase oxygen content, in the case of $Mo_2TiC_2T_x$ and $Mo_2Ti_2C_3T_x$ MXenes.[38] Such cationic organic molecules can be readily intercalated between the negatively charged surfaces of multilayer MXene that allows effective delamination of 2D flakes. Alkalization of MXene using base solution is reported to transform F-groups into OH-groups in $Ti_3C_2T_x$.[39] Hydrazine treatment of $Ti_3C_2T_x$ MXene has been reported to decrease the water contents and OH group, which was beneficial for enhanced volumetric capacitance.[40]

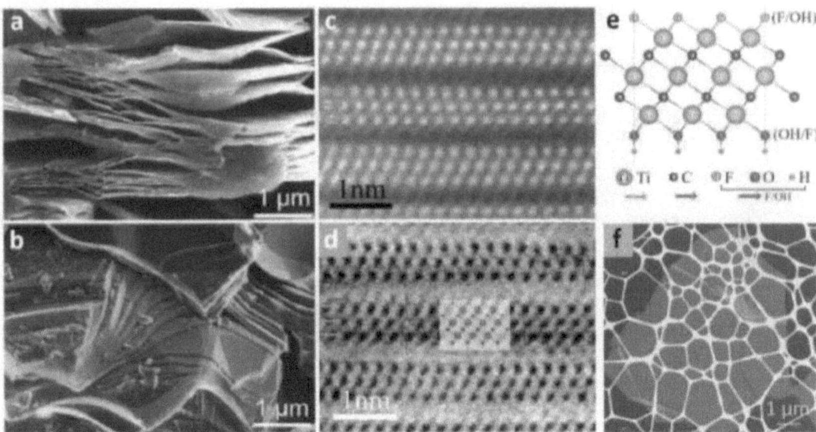

Figure 2.12. a-b) Scanning electron microscopy (SEM) images of HF route $Ti_3C_2T_x$ with accordion-like structure (a) and HCl-LiF route $Ti_3C_2T_x$ with a more compact structure (b). Reproduced with permission.[37] Copyright 2016, PCCP Owner Societies. **c)** high-angle annular dark-field (HAADF) image of multilayer $Ti_3C_2T_x$, showing ABAB stacking of adjacent two $Ti_3C_2T_x$ monolayers as inherited from the parent Ti_3AlC_2 MAX phase. **d)** Annular bright-field (ABF) image combined to HAADF, revealing the surface functional groups of $Ti_3C_2T_x$. **e)** Schematic description of the proposed atomic structure of $Ti_3C_2T_x$, where the surface functional groups locate the hollow site of three neighbor carbon atoms, creating ABCABC stacking of atomic layers. Reproduced with permission.[41] Copyright 2015, American Chemical Society. **f)** Annular dark-field scanning transmission electron microscopy (ADF STEM) image of delaminated $Ti_3C_2T_x$ nanosheet. Reproduced with permission.[29] Copyright 2017, WILEY-VCH Verlag GmbH & Co. KGaA, Weinheim.

2.4.2 Tunable Interlayer Spacing

MXenes also have tunable interlayer spacing by controlling intercalated species between the nanosheets. Once cation species are intercalated, $Ti_3C_2T_x$ MXene shows hygroscopic property and reversible humidity dependent behavior with interlayer spacing of 12.5-15.5 Å (including one or two intercalated water layers).[20] Organic molecules can also be intercalated between MXene nanosheets. For instance, $Ti_3C_2T_x$ with pillared structure has been reported using cetyltrimethylammonium bromide (CTAB) or stearyltrimethylammonium bromide (STAB) with interlayer spacing of 22.3 or 27.1 Å, respectively.[42] A series of Alkylammonium cations have been intercalated and demonstrated tunable interlayer spacing between 14.7-38.0 Å.[43] Tetrabutylammonium cation (TBA^+) intercalated in Mo_2CT_x is reported to have interlayer spacing of 16.9 – 18.9 Å depending on the degree of wetness.[19, 44] Such tunable interlayer spacing can be useful for gas sensing and gas separation applications.

2.4.3 Electronic Band Structure

A number of theoretical studies have been conducted to identify the electronic band structures and interesting properties of MXenes from their large family.[45] Most of the functionalized MXenes are predicted to have metallic/semi-metallic band structures, while a few MXene systems are expected to be semiconducting. Sc_2CT_2 (T = O, OH, F), Ti_2CO_2, Zr_2CO_2, Hf_2CO_2 are reported to have band gaps between 0.24 to 1.8 eV.[7] The reported band structure and band gaps are shown in **Figure 2.13** and **Table 2.1**, respectively. Only $Sc_2C(OH)_2$ is expected to have direct band gap while the others are predicted to have indirect band gaps among those 6 MXenes. Interestingly, the indirect to direct band gap transition can be induced by the presence of biaxial strain of 4, 10, and 14% for Ti_2CO_2, Zr_2CO_2, and Hf_2CO_2, respectively.[46] The transition from bare MXenes (metallic) to functionalized MXenes is accompanied by a band gap opening due to newly created states below the Fermi level by strong band hybridization between M $3d$ orbital and C $2p$ or O $2p$ orbitals.[47] This can be explained by the lower electronegativity of transition metals compared to the functional groups and carbon atoms. The OH and F groups can receive one electron, while the O group can receive two electrons from the transition metal atoms. In addition, Ti, Zr, and Hf locate at the same column in the periodic table with the same outer-shell electronic configuration (two electrons in s- and d-orbitals), hence similar metallic to semiconducting character transition trends have been discovered in the corresponding M_2C MXene forms after O group functionalization. The band gap of M_2CO_2 (M = Ti, Zr, Hf) MXenes increase following the metal (M) atomic number, due to the weakening in the metal electronegativity.[48] For the $Ti_{n+1}C_nO_2$ (n≥2) MXenes, the contribution of O $2p$ orbital diminishes as n increases and such MXenes are no longer semiconducting.[49] Mo_2C and Cr_2C are reported as semiconductors with OH, F, and Cl functional groups, while several double transition metal MXenes are semiconductors upon O termination. (See **Table 2.1**)

The experimental realization of thin-film transistor (TFT) using MXene as semiconducting channel is yet to be achieved, however, metallic MXenes can be used as contact materials where their hydrophilic nature can allow easy integration with various semiconducting materials including oxides, polymers, 2D transition metal dichalcogenides (TMDs), and quantum dots.

Few MXene systems with two different metal atoms together can create an ordered structure instead of randomly distributed solid-solution. These are categorized as ordered double transition metal MXenes, with the general formula of $M'_2M''C_2$ and $M'_2M''_2C_3$, where M' forms the outer surfaces and M'' form the central layers of the structure.[26] For example, $Mo_2TiC_2T_x$ has Mo atoms at the surface layers while Ti atoms are located in the central layers. Such ordered structure is known to affect the electronic properties of MXenes. For example, $Ti_3C_2T_x$ is metallic while $Mo_2TiC_2T_x$ is semiconducting.[38] Moreover, several oxygen-terminated double transition metal MXenes, $M'_2M''C_2O_2$ and $M'_2M''_2C_3O_2$ (M' = Mo, W; M'' = Ti, Zr, Hf), have been predicted to be 2D topological insulators and topological semimetals, respectively.[50] A sizable topological gaps of 0.1-0.2 eV has been reported in $Mo_2M''C_2O_2$ (M''=Ti, Zr, Hf) system, which is even large enough to realize quantum spin Hall effect at room temperature.[51] The energetically preferred oxygen-terminated surface can protect MXenes from oxidation under air ambient. Besides, M_2CO_2 (M = W, Mo, Cr) MXenes are also topological insulators.[52] Some of those topologically non-trivial MXenes have already been experimentally synthesized, hence these MXenes can be good candidate materials for electronic, spintronic devices, and topological superconductivity research.

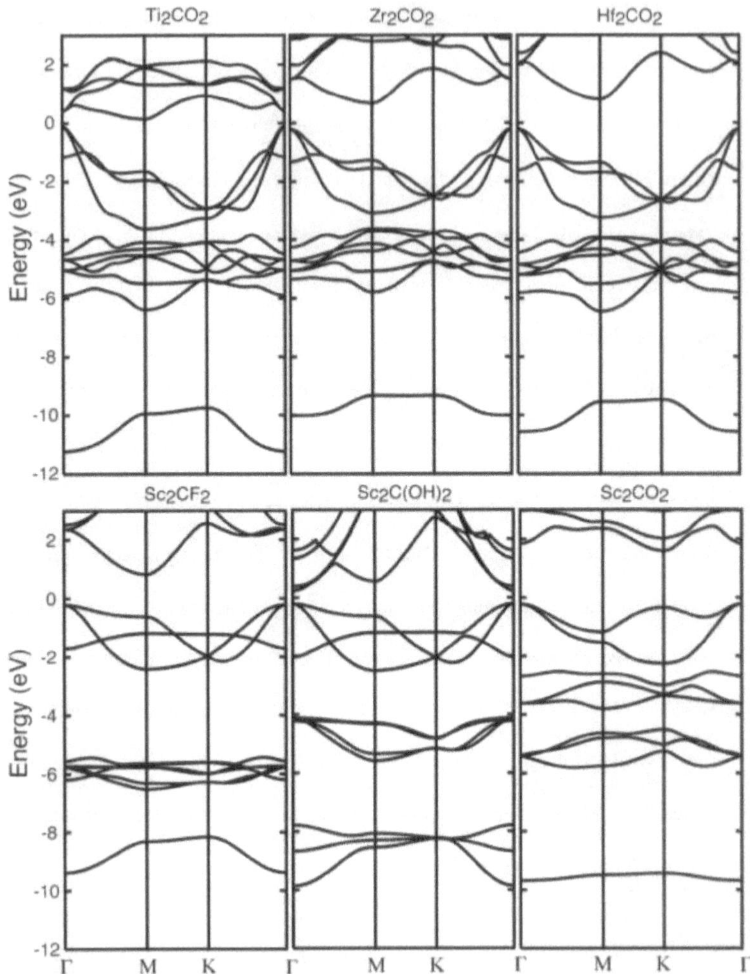

Figure 2.13. Band structure of semiconducting M_2CT_2 MXene systems. The Fermi energy is at zero. The transition metal element and surface functional groups significantly affect the electronic band structure of MXenes. Reproduced with permission.[7] Copyright 2013, WILEY-VCH Verlag GmbH & Co. KGaA, Weinheim.

Table 2.1. Calculated band gaps [eV] of functionalized semiconducting MXenes based on conventional (PBE) and hybrid (HSE06) functionals. The latter is more accurate in band gap prediction. Direct band gap MXenes are marked, unless otherwise indirect band gap. Specified properties are remarked such as antiferromagnetic (AFM), ferromagnetic (FM), and topological insulator (TI).

MXene	Termination	PBE	HSE06	Remark
Sc_2C	O	$1.8^{[7]}$, $1.84^{[53]}$, $1.86^{[54]}$	$2.90^{[55]}$, $2.92^{[56]}$, $3.01^{[54]}$	
	F	$1.0^{[54, 57]}$, $1.03^{[7, 53]}$, $1.05^{[58]}$	$1.64^{[58]}$, $1.84^{[55]}$, $1.88^{[54]}$	
	OH	$0.34^{[54]}$, $0.44^{[53]}$, $0.45^{[7]}$, $0.71^{[57]}$	$0.71^{[54]}$, $0.74^{[55]}$	Direct band gap
	Cl	$0.88^{[55]}$	$1.64^{[55]}$	
Ti_2C	O	$0.17^{[57]}$, $0.24^{[7, 49]}$, $0.33^{[48]}$	$0.78^{[59]}$, $0.88^{[49]}$, $0.92^{[60]}$	
Zr_2C	O	$0.66^{[57]}$, $0.88^{[7]}$, $0.95^{[48]}$	$1.54^{[60]}$	
Hf_2C	O	$0.8^{[57]}$, $1.00^{[7, 48]}$	$1.657^{[61]}$, $1.75^{[60]}$	
V_2C	F	$0.56^{[62]}$		AFM
	OH	$0.44^{[62]}$		AFM
Cr_2C	O			TI
	F	$0.22^{[57]}$	$3.15^{[63]}$, $3.49^{[64]}$	AFM
	OH	$0.03^{[57]}$	$1.39^{[63]}$, $1.76^{[64]}$	AFM
	Cl	$0.15^{[57]}$	$2.56^{[64]}$	AFM
Mo_2C	O			TI
	F	$0.25^{[57]}$		
	OH	$0.1^{[57]}$		
	Cl	$0.15^{[57]}$		
W_2C	O	$0.194^{[52]}$	$0.472^{[52]}$	TI
$(Mo_{2/3}Sc_{1/3})_2C$	O	$0.04^{[65]}$	$0.58^{[65]}$	Piezoelectric
$(Mo_{2/3}Y_{1/3})_2C$	O	$0.45^{[65]}$	$1.23^{[65]}$	Piezoelectric
$(W_{2/3}Sc_{1/3})_2C$	O	$0.675^{[65]}$	$1.3^{[65]}$	Piezoelectric
$(W_{2/3}Y_{1/3})_2C$	O	$0.625^{[65]}$	$1.3^{[65]}$	Piezoelectric
$Mo_{1.33}C$	$O_{2/3}F_{1/3}$	$0.5^{[66]}$		

Hf_3C_2	O		0.155[67]	
Hf_2MnC_2	O	0.238[68]		FM
	F	1.027[68]		AFM
Hf_2VC_2	F	0.4[69]	0.9[69]	Multiferroic (type-II)
Mo_2TiC_2	O	0.041[50], 0.052[51]	0.119[50], 0.125[51]	TI
Mo_2ZrC_2	O	0.069[50], 0.087[51]	0.125[50], 0.147[51]	TI
Mo_2HfC_2	O	0.153[50], 0.213[51]	0.238[50], 0.301[51]	TI
W_2TiC_2	O	0.136[50]	0.290[50]	TI
W_2ZrC_2	O	0.170[50]	0.280[50]	TI
W_2HfC_2	O	0.285[50]	0.409[50]	TI
Cr_2TiC_2	F		1.35[63]	AFM
	OH		0.85[63]	AFM, Direct Band gap

2.4.4 Work Function

Locating at the surface of MXenes, the functional groups are directly related to the work function, which is critical for electronic device applications. Khazaei *et al.* have reported OH-terminated MXenes tend to have ultralow work functions while other functional groups always induce large work function.[70] The theoretical studies discovered a sufficient wide work function range for the functionalized MXenes, between ca. 1.6 and ca. 8.0 eV from $Sc_2C(OH)_2$ and Cr_2CO_2, respectively.[70, 71] (**Figure 2.14 (a)**). Such work function range is even wider than conventional metals (dashed lines in **Figure 2.14 (a)**), which suggests the promise of functionalized MXenes as contact materials in electronic devices. There is an interesting negative correlation between the work function of O-terminated MXene and OH-terminated MXene (**Figure 2.14 (b)**), where the extremely large modulation of work function can be utilized in novel electronic/sensing devices. Surface terminations not only draw electrons from MXene core but also change the surface dipole moment, hence affect the work function of the host MXenes.[70, 72] (**Figure 2.14 (c)**) Substitutionally doped MXene with alkali and/or alkali earth metals can further reduce their work functions down to ca. 1.2 eV.[72] Moreover, OH-terminated MXenes have been reported to have nearly free electron (NFE) states which are partially occupied and located near Fermi level.[73] (**Figure 2.14 (d) and (e)**) This implies that the electron transport in OH-terminated MXenes can be free of nuclear scattering, hence it can be useful for electrical applications. The NFE states, however, disappear upon heterojunction formation with other 2D materials.

Liu *et al.* have proposed MXenes as Schottky barrier-free contact materials for semiconducting TMDs.[71] For example, O-terminated MXenes are suitable for Schottky-barrier-free hole injection for p-type WSe_2. The proposed concept is not only based on the high work function of O-terminated MXenes but also the weak van der Waals interaction at metal-semiconductor interface, which can release the Fermi-level pinning effect.[74] Similarly, the OH-terminated MXenes with low work functions are suitable contact for n-type MoS_2.

In metal-semiconductor junctions, the Mott-Schottky rule is generally used to predict or explain the contact conditions. Depending on the difference between the work function (metal) and electron affinity or ionization energy (semiconductors), either Ohmic contact or Schottky junctions with certain barrier height can be achieved. However, the experimental results often deviate from the fundamental theory, which is mainly due to the interfacial defects and Fermi-level pinning effect. For example, high-energy metal deposition methods such as e-beam evaporation can damage underlying material and create gap states. Recently, it has been demonstrated that transferred metal electrodes can realize van der Waals metal-semiconductor junctions following the Mott-Schottky rule.[75] In their work, the MoS_2 transistors with various transferred metal electrodes have shown completely different transistor operating behaviors from the cases using the common high-energy deposited metals contacts. Therefore, the low-energy metal deposition method is important, and the solution processed MXenes can be one of the practical options.

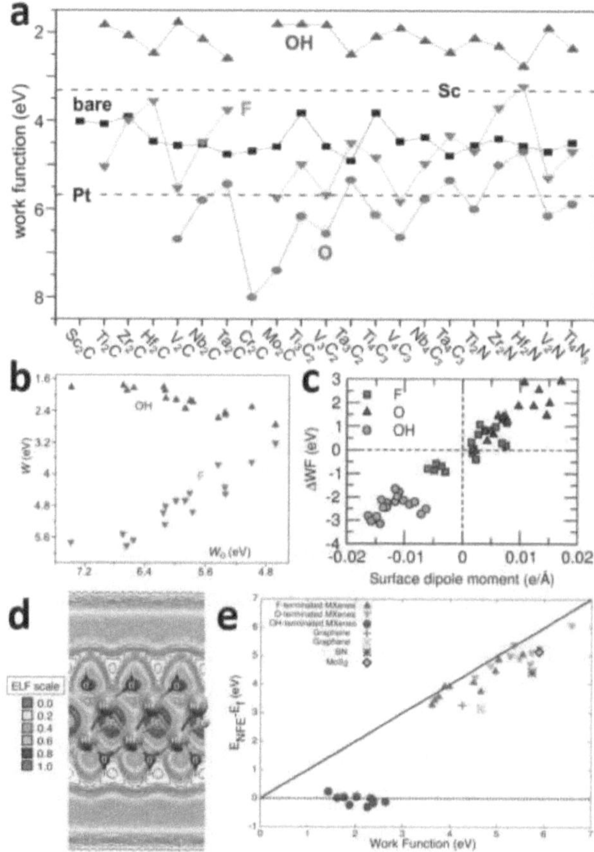

Figure 2.14. a) Calculated work functions of bare and terminated MXenes in comparison with Sc and Pt. **b)** Work functions of OH- and F- functionalized MXenes as a function of O- functionalized MXenes. Reproduced with permission.[71] Copyright 2016, American Chemical Society. **c)** Surface functionalization induced changes in work function of MXenes with respect to the surface dipole moment. Reproduced with permission.[72] Copyright 2017, The Royal Society of Chemistry. **d)** Electron localization function (ELF) contour plot of $Hf_2C(OH)_2$ showing the NFE states outside of MXene. **e)** The relative energy position of the lowest NFE state with respect to the Fermi level (E_f) as a function of work function for functionalized MXenes in comparison with graphene, MoS_2, and BN. The solid line is the vacuum level. Reproduced with permission.[73] Copyright 2015, American Physical Society.

Chapter 3

Oxide Thin Film electronics with all-MXene Electrical Contacts

3.1 Introduction

Delaminated transition metal (M) carbides and/or nitrides (X), known as MXene, belong to the rising two-dimensional (2D) materials family that exhibit great potential in energy storage,[35] catalysis,[16] and electromagnetic shielding applications.[14] MXenes can be obtained by chemical etching of the layered precursor MAX phases, where the acidic solution attacks the "A-element" in metallic M-A bonds.[1] This process results in spontaneous formation of surface functional groups (including –OH, =O, and –F).[2] These surface groups make the nanoflakes hydrophilic and hence improve their dispersion in aqueous solutions. Theoretical studies have predicted that surface termination groups play a key role in controlling the properties of various MXene systems.[7, 45] Combining the unique hydrophilic surface and metallic conductivity, MXene based supercapacitors have shown impressive progress, including operating at an ultrafast cycling rate of $\sim 10^5$ mV s^{-1}, and very high volumetric capacitance of 1500 F cm^{-3}.[35] The recent investigation of MXene has largely focused on their electrochemical properties,[1] while their potential use in electronics has been significantly less explored. Recently, we have shown a relatively large thermoelectric power factor of 3.09×10^{-4} W m^{-1} K^{-2} for Mo_2TiC_2, which is stable up to 800 K.[19] Xu et al. reported that mechanically exfoliated MXene (Ti_2C) can be used as current emitters for 2D molybdenum sulfide (MoS_2) and tungsten selenide (WSe_2) nanoflakes, which shows small Schottky barrier heights of 0.19 and 0.23 eV, respectively.[76] Kang et al. demonstrated photodetecting behavior of Schottky junction between drop-casted MXene (Ti_3C_2) and n-Si.[77] Xu et al. reported MXene (Ti_3C_2) based ion-sensitive field-effect transistor for biosensing applications by soft-lithography method.[78] However, the current stage of MXene based electronic devices still requires extra contact materials, such as Si^{++} gate or additional metal electrodes, even though large-scale MXene thin film processes (spin or spray coating) have already been demonstrated.[79, 80] The unique combination of metallic conductance and hydrophilic surface suggests MXene can potentially be a good contact material in electrical switching devices (both gate and source/drain contacts). The metallic conductivity of MXene can allow fast response of the device operation in both static and dynamic tests. In addition, thanks to its excellent electrochemical properties, devices that combine electronic and electrochemical functions of MXenes may be fabricated. Another attractive feature is that the hydrophilic MXene surface allows facile growth of metal oxide dielectrics by atomic layer deposition (ALD) without additional treatment. This is in sharp contrast to the difficult growth of ALD oxide dielectrics on the hydrophobic surfaces of graphene and/or other 2D materials.[81, 82]

In this work, for the first time, we demonstrate that n-type and p-type oxide thin-film transistors (TFTs) can be fabricated using large-area MXene (Ti_3C_2) electrical contacts (gate, source, and drain). The key to this result is that the work function of MXene (Ti_3C_2) matches well with the band edges of both zinc oxide (ZnO) and tin monoxide (SnO), leading to good Ohmic contact formation. Spray-coating was used to form the MXene films because it is a cost-effective technique that has already been successfully demonstrated in thin film electronics, and because it is compatible with large-area processing methods such as inkjet printing and roll-to-roll processing. Individual TFTs and complementary metal oxide semiconductor (CMOS) inverters with all-MXene contacts and reasonably good performance have been achieved.

3.2 Materials Characterization

Basic properties of the spray-coated MXene film (Ti_3C_2) are shown in **Figure 3.1.** The sheet resistance (R_S) was measured by four-point method. The I-V curve is shown in **Figure 3.1(a)**, and the measurement schematic is shown as top inset. The R_S is calculated to be 41.8 Ω/sq, based on the fitted slope of I-V curve and relative dimension of the sample and tip spacing. The optical image of a spray-coated MXene film on glass substrate is shown as the bottom inset to **Figure 3.1(a)**, where the MXene film shows semi-transparent dark greenish color. The metallic behavior of the MXene film is confirmed by the monotonic increase of resistivity with temperature over the range from 300 to 800 K (**Figure 3.1(b)**), which agrees with previous reports. [19, 83] The work function (Φ) of MXene (Ti_3C_2) is of great importance, since it will be used as contact material to both n-(ZnO channel) and p-type (SnO channel) transistors. Photoelectron spectroscopy in air (PESA) technique was used to measure the work function of the contact material (MXene), conduction band (E_C, or electron affinity) of n-type ZnO, and valence band (E_V, or ionization energy) of SnO, as shown in **Figure 3.2(c)**. The Φ_{MXene}, $E_{C,ZnO}$, and $E_{V,SnO}$ are measured to be 4.60, 4.58, and 4.65 eV, respectively. These values indicate negligible barrier exists for carrier injection into the corresponding transistor channel, since the Φ of n-type (p-type) semiconductor is deeper (shallower) than the E_C (E_V) level. The surface XPS study suggests that our MXene films have majority =O termination with minor portion of –OH and –F surface functional groups. The XRD patterns of Ti_3AlC_2 MAX powder and spray-coated Ti_3C_2 MXene film are shown in **Figure 3.1(d)**. The XRD pattern of the MAX powder matches well with literature reports.[2] The c-lattice parameter (c-LP) is calculated to be 18.6 Å, according to the ($00l$) peak positions. The XRD pattern of the MXene film is featureless but shows only intensive (002) peak, which indicates successful delamination of 2D MXene flakes and highly c-oriented restacking nature of MXene film. The c-LP is calculated as 27.6 Å, which is largely expanded compared to the precursor counterpart. The increased c-LP suggests presence of surface functional groups, intercalated lithium cations and water molecules between MXene nanoflakes.[3] The Raman spectra of Ti_3AlC_2 (MAX) powder and Ti_3C_2 (MXene) film are shown in **Figure 3.1(e)**. The distinct sharp peaks of Ti_3AlC_2 shift and transform to broader spectrum of Ti_3C_2 after chemical etching of the aluminum layer, which agrees with the literature. [19, 29, 84] The broadened Raman bands are due to the co-existence of multiple surface functional groups that can affect phonon dispersion of MXenes.[85] The SEM image of the MXene film is shown in **Figure 3.1(f)**. The lateral size of single flake is around 1.5 μm, and the flakes are uniformly coated on the substrate with good coverage. **Figure 3.1(g)** shows the atomic force microscopy (AFM) image of ultra-thin MXene film (separately prepared, only for measuring flake thickness). The thickness of the single nanoflake is measured to be 1.6 nm (shown as inset), which agrees with previous literature reports.[29] The AFM images of the as-sprayed and HfO_2-coated MXene films are shown in **Figure 3.1(h) and (i)**, respectively. The neighboring MXene flakes overlap and form conductive 2D nanoflake network. The ALD processed HfO_2 dielectric is uniformly coated on the MXene flakes. The root mean square (rms) roughness for the as-sprayed MXene film is 3.69 nm, which increases to 8.22 nm after coating HfO_2. The rms roughness of the HfO_2 on MXene film is higher than typical ALD HfO_2 films deposited on other conducting materials (~5 nm),[86] which indicates that there is room for improvement with further process optimization.

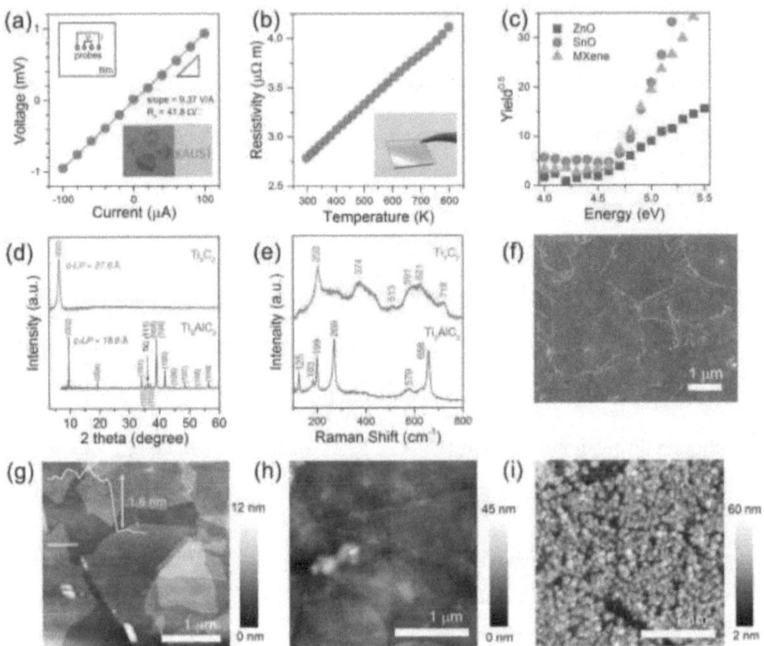

Figure 3.1. Characterizations of spray-coated Ti_3C_2 MXene films. a) I-V curve obtained by the four-point method, the insets show the measurement setup and optical image. b) Resistivity vs temperature curve of spray-coated MXene film, inset shows the optical image. c) PESA spectra of the MXene, ZnO, and SnO films used for work function extraction. d-e) XRD patterns (d) and Raman spectra (e) of the Ti_3AlC_2 MAX (powder) and Ti_3C_2 MXene film. f) SEM image of the MXene film. g-i) AFM images of MXene films (g, h) and HfO_2 coated MXene film (i). Reprinted with permission.[30] Copyright 2018 WILEY-VCH Verlag GmbH & Co. KGaA, Weinheim.

3.3 Device Performance

3.3.1 Device Fabrication Process

The schematic process flow for MXene (Ti_3C_2) film deposition and device fabrication is shown in **Figure 3.2**. The MAX phase (Ti_3AlC_2 powder) was chemically etched by hydrochloric acid and lithium fluoride mixture solution (**Figure 3.2(a)**).[3, 29] Sufficient reaction time was used to ensure that the aluminum atoms were removed at the end of this step (**Figure 3.2(b)**). The delaminated MXene flakes were collected by centrifugation after repeating the wash cycle with deionized water and mild hand shaking (**Figure 3.2(c)**). The as-prepared MXene solution with a typical concentration of $1.0 - 1.5$ mg/ml was used for spray-coating MXene films on glass substrates (**Figure 3.2(d)**). The MXene film was then loaded in the ALD chamber to deposit the hafnium oxide (HfO_2) dielectric layer (**Figure 3.2(e)**). The n-type semiconducting ZnO film was grown directly on top of HfO_2 in the same ALD chamber, while the p-type SnO film was grown in a separate magnetron sputtering chamber (**Figure 3.2(f)**). The whole device was completed by spray-coating another MXene source/drain contact on top of the patterned semiconducting layers (**Figure 3.2(g)**). The ZnO film was patterned by wet-etching, while the SnO and MXene films were patterned by the lift-off technique (See Experimental for details).

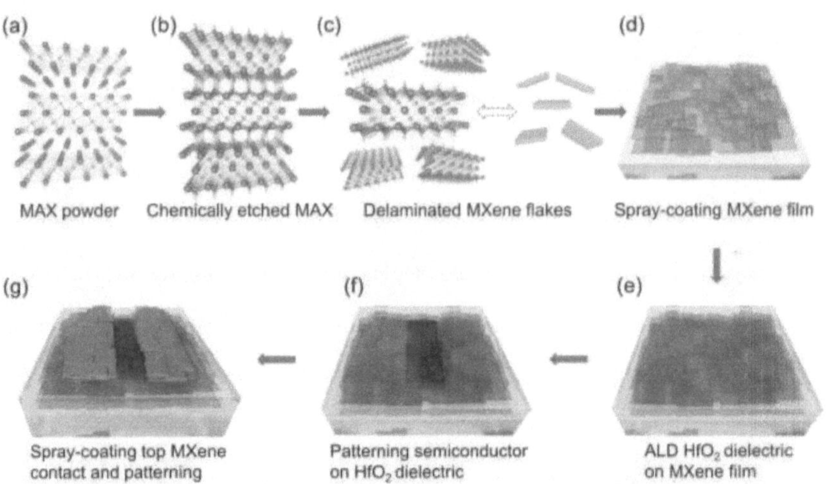

Figure 3.2. The schematic process flow for MXene (Ti$_3$C$_2$) film and device fabrication. a-c) Crystal structure of (a) MAX (Ti$_3$AlC$_2$), (b) chemically-etched MAX, and (c) delaminated MXene flakes. d-g) The fabrication process of TFT devices with all-MXene contacts (d) spray-coating MXene film on glass substrate, (e) HfO$_2$ dielectric growth by ALD on MXene, (f) semiconductor patterning, (g) spray-coating of top MXene contacts and patterning. Reprinted with permission.[30] Copyright 2018 WILEY-VCH Verlag GmbH & Co. KGaA, Weinheim.

3.3.2 Oxide Thin Film Transistors

The performance of n- and p-type TFTs with all-MXene contacts is shown in **Figure 3.3**. The schematic structure of the TFT device is shown in **Figure 3.3(a)**, using typical bottom-gate, staggered structure.[87] Detailed information on the n-type ZnO and p-type SnO semiconductors can be found in our previous reports.[86, 88, 89] The channel dimension used for both n- and p-type TFTs is 100×100 μm^2. The capacitance-voltage (C-V) curves of MXene/HfO$_2$/MXene metal/insulator/metal (MIM) capacitor (diameter: 200 μm) are shown in **Figure 3.3(b)**. The measured areal capacitance (C$_{ox}$) is 1.04×10^{-7} F cm^{-2}, calculated dielectric constant of HfO_2 is 16.5, and measured leakage current density is 5×10^{-6} A cm^{-2}. The output curves of p- and n-TFT are shown in **Figure 3.3(c) and (d)**, respectively. The output curve of n-TFTs shows strong saturation compared with p-TFTs, indicating the formation of sharp pinch-off region in the channel near the drain electrode.[90] The absence of current crowding effect in the low V_{DS} range of both output curves suggests that good contact exists between MXene source/drain and both semiconductor channels. Previously, titanium was reported to show good contact with both p-SnO[89, 91-93] and n-ZnO[88, 94], and it seems that Ti_3C_2 MXene contacts behave similarly to the titanium electrodes. The measured work function values by PESA also suggest that good contact should exist between MXene (4.60 eV) and either n-type ZnO (4.58 eV) or p-type SnO (4.65 eV) semiconductor. The transfer curves of p- and n-type TFTs are shown in **Figure 3.3(e) and (f)**, respectively. During the transfer curve measurement, small source to drain voltage (V_{DS}) of -1 and 1 V were applied to induce hole (p-type) and electron (n-type) current in the corresponding channels. Under such V_{DS}, both devices operate in the linear regime. Both devices show small leakage current (I_{GS}, below 10^{-10} A), indicating the effect of gate leakage current on mobility extraction is negligible. The p-type TFT shows relatively large hysteresis in the dual sweep loop, where the threshold voltage shift is ~3.3 V. Similar behaviors of p-type SnO TFTs have commonly been observed, which is attributed to the presence of charged trapping centers at the dielectric/semiconductor interface.[87] The voltage hysteresis in n-type device is smaller than p-type TFT, and is about 0.5 V. However, this hysteresis value is larger than previously reported for ZnO TFTs (~0 V), [86, 88] which indicates that more defects might have formed at the dielectric/semiconductor interface during ZnO growth on the rough HfO_2 surface.

Systematic optimization of large-area spray-coated MXene films are under way to improve TFT device performance. The switching ratios (I_{on}/I_{off}) are calculated to be 1.1×10^3 and 3.6×10^6 for p- and n-type TFTs, respectively. The subthreshold swing (SS) of both types of TFTs are extracted from the inverse of the maximum slope in the logarithmic-scale transfer curve. The calculated SS values are 2.51 and 0.23 V dec^{-1} for p- and n-type TFTs, respectively. The field-effect mobility (μ_{FE}) of p- and n-type TFTs are calculated based on the linear plot of the transfer curves, as shown in **Figure 3.3(g) and (h)**, respectively. The μ_{FE} of the p-type and n-type TFTs are calculated to be 2.01 and 2.61 cm^2 V^{-1} s^{-1}, respectively. Although somewhat higher μ_{FE} value can be extracted from the transfer curve of the n-type TFT, it still shows lower current in the output curve. This might relate to the early pinch-off observed in the ZnO channel; alternatively, the ALD ZnO channel might be more defective due to the surface roughness of the HfO$_2$ dielectric layer. The threshold voltage (V_{th}) values of the TFTs can also be extracted from these linear plots, and are determined to be -1.09 V and 4.25 V for p- and n-type TFTs, respectively.

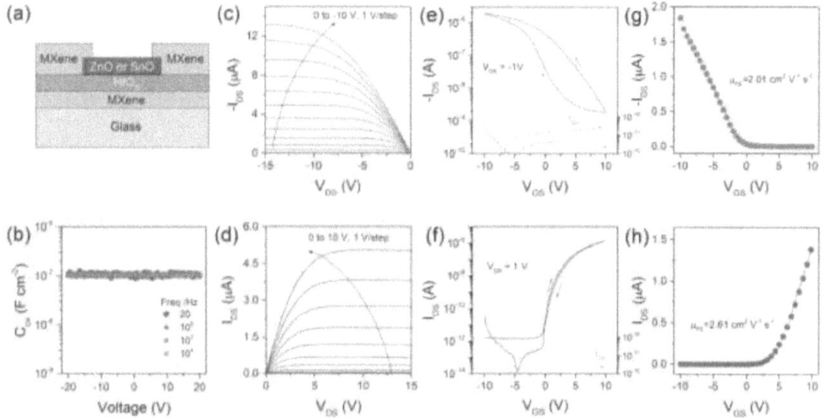

Figure 3.3. Performance of TFTs with all-MXene (Ti₃C₂) contacts. a) Schematic structure of the device. b) Capacitance-voltage curve of MXene/HfO₂/MXene MIM structures. c,e,g) Output (c), transfer (e) characteristic, and (g) mobility extraction for the p-type TFT. d,f,h) Output (d), transfer (f) characteristic, and (h) mobility extraction for the n-type TFT. Reprinted with permission.[30] Copyright 2018 WILEY-VCH Verlag GmbH & Co. KGaA, Weinheim.

3.3.3 Complementary Metal–Oxide–Semiconductor (CMOS) Inverter

The balanced performance of the n- and p-type TFTs indicates a well-performing CMOS inverter with MXene contacts can be fabricated. The performance of such CMOS inverter with all-MXene contacts is shown in **Figure 3.4**. The schematic circuit diagram is shown in **Figure 3.4(a)**. Two TFTs share a common gate electrode (the input terminal). The source electrode of the p-type TFT is used as the V_{DD} (supply voltage) terminal. The source electrode of n-type TFT is used as the V_{SS} terminal, which is grounded. The output terminal is built by connecting the drain electrodes of both TFTs. The n-type TFT channel ratio is intentionally increased to better balance the output current (or channel resistance). The channel widths are 250 and 100 μm for n- and p-type TFTs, respectively, while the channel lengths are maintained at 100 μm. When a small V_{input} is applied, the p-channel is turned on, and the n-channel off, thus the V_{output} becomes equal to V_{DD} (high state). When the V_{input} voltage increases, the p-channel is gradually turned off and the n-channel is gradually turned on, thus the V_{output} drops sharply (transition state). Finally, if sufficiently large V_{input} is applied, the n-channel is turned on, and the V_{output} equals to V_{SS} (low state). The voltage transfer curve (VTC) of this CMOS inverter is shown in **Figure 3.4(b)**, where a V_{DD} of 10 V was applied. Slight hysteresis is observed from the dual-direction scanning, and the V_M (switching threshold voltage, V_{input} value when $V_{input} = V_{output}$) is found to be 5.02 and 5.27 V in the forward and backward scanning directions, respectively. The maximum V_M variation (0.25 V) is 5% of $V_{DD}/2$, indicating that the logic switching is stable regardless of the scan direction. The VTC characteristic exhibits rapid output voltage transition, and the voltage gain value was calculated from $\partial V_{output}/\partial V_{input}$, which is plotted in **Figure 3.4(c)**. The peak gain value is determined to be 80, and the CMOS inverter exhibits a logic swing range (V_L) of **8.78 V**, which is **87.8%** of the ideal case (when gain is infinity, logic swing is equal to V_{DD}). The peak I_{DD} value is 1.1 μA, which corresponds to maximum power consumption of 11 μW. The static power consumption is estimated to be 0.28 μW, by $[V_{DD}\times(I_{DD,H}+I_{DD,L})/2]$, where $I_{DD,H}$ and $I_{DD,L}$ is the I_{DD} values at high ($V_{input}=0$ V) and low ($V_{input}=10$ V) states, respectively. The butterfly VTC plot is shown in **Figure 3.4(d)**, where another VTC is plotted in mirror. The noise margin can be estimated by nesting the largest possible rectangular area inside the butterfly plot. The maximum equal criterion (MEC) method is used to estimate the noise margin (NM).[95] The NM of this CMOS inverter is calculated to be 3.54 V, which is 70.8% of the ideal value ($V_{DD}/2$), and is enough for most of the static logic applications. The dynamic response of the CMOS inverter is shown in **Figure 3.4(e)**, where a 100 Hz square waveform is used as input. The output signal is inverted accordingly, indicating the CMOS inverter works properly at such frequency. Detailed investigation is performed on the output signal (**Figure 3.4(f)**), where the output signal rise and fall delay are measured between 10% and 90% of the output signal range. The signal rise and fall delay are 63 and 143 μs, respectively. Such delay time indicates that the CMOS inverters can work for kHz range applications. We believe that better CMOS performance could be achieved with further optimization, including optimizing the channel length, parasitic capacitance, and the mobility of channels. However, in this first ever demonstration of large-area, all-MXene-contacted CMOS inverters, the large gain value, small static power consumption, wide logic swing window, excellent noise margin, and 100 Hz dynamic operation show the potential for metallic Ti_3C_2 MXene contacts in electronic applications. The advantage of Ti_3C_2 MXene compared with other solution-processed 2D materials for electronic applications are summarized in **Table 3.1**. Basically, the high conductivity and ALD-friendly hydrophilic surface make it likely that Ti_3C_2 MXene may offer some advantages in electronic device applications.

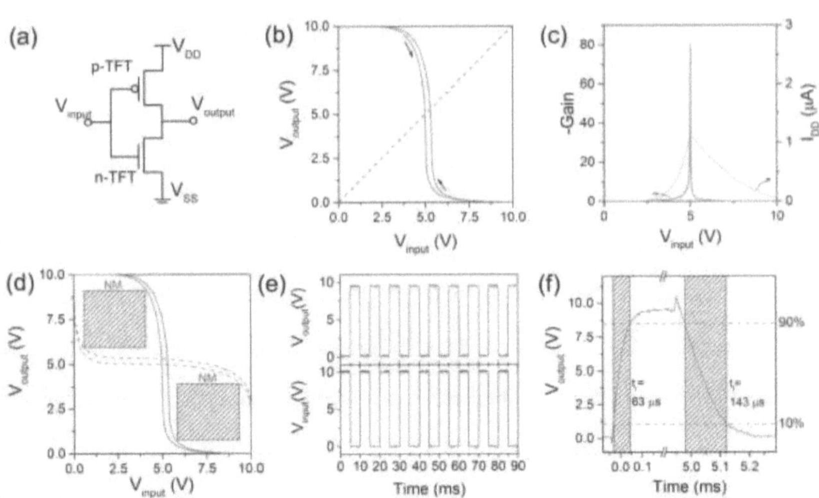

Figure 3.4. Performance of CMOS inverter fabricated using all-MXene contacts. a) Schematic circuit diagram of the CMOS inverter. b,c) VTC (b), gain and I_{DD} (c) of the CMOS inverter. d) Noise margin extraction using the MEC method from the butterfly plot. e,f) Dynamic response (e) to a 100 Hz square input waveform, (f) extraction of the output signal delay during the rise and fall cycles. Reprinted with permission.[30] Copyright 2018 WILEY-VCH Verlag GmbH & Co. KGaA, Weinheim.

Table 3.1. Solution-processed two-dimensional materials for electrode devices.

Materials[a]	Conductive?[b]	Hydrophilic?	Aqueous dispersion?	ALD growth?	Method[c]	σ_{dc}[d]	R_S[d]	T@550 nm[d]	Ref.
						S/cm	Ω/sq	%	
Ti_3C_2 MXene	O	O	O	O	S	3599	41.8	42	This work
					VF[e]	10166	0.196	0	This work
					SP	5736	422	93	[29]
					SP[e]	9880	11	29	[29]
					S	-	8000	90	[80]
					S[e]	2200	500	40	[80]
Graphene	O	X	X	X	VF	6	2.7×10^5	92	[96]
					VF[e]	150	7.0×10^3	36	[96]
					D	58	-	97.7	[97]
					D	124	-	95.4	[97]
					D	150	-	93.1	[97]
GO	X	O	O	O		-			
rGO/CCG	△	X	X	△	S	-	2.0×10^7	96	[98]
					VF[e]	72	-	0	[98]
					LB	-	10492	80	[99]
					VF	-	1.0×10^8	95	[100]
					VF	-	4.3×10^4	63	[100]
					VF	-	5.0×10^5	83	[101]

a) Materials: GO: graphene oxide, rGO: reduced graphene oxide (including chemically modified graphene). b) Evaluation of the performance, O: good, △: moderate, X: poor. c) Method for film preparation. S: spray coating, SP: spin casting, VF: vacuum filtration, LB: Langmuir-Blodgett assembly, D: drop casting. d) In air at room temperature. e) Bulk transport, unless otherwise percolative.

Reprinted with permission.[30] Copyright 2018 WILEY-VCH Verlag GmbH & Co. KGaA, Weinheim.

3.4 Conclusions

To summarize, we have successfully demonstrated n- and p-type oxide thin film transistors using all-MXene (Ti3C2) electrical contacts (gate, source, and drain). Zinc oxide and tin monoxide are used as n- and p-semiconducting channels, respectively. Work function measurement indicates negligible band offsets between MXene and both type of semiconducting channels. The n- and p-type transistors with all-MXene contacts showed balanced performance, including field-effect mobility of 2.61 and 2.01 cm^2 V^{-1} s^{-1}, threshold voltage of 4.25 and -1.09 V, and switching ratio of 3.6×10^6 and 1.1×10^3, respectively. In addition, CMOS inverters with all-MXene contacts have been demonstrated with very promising performance. At a supply voltage of 10 V, the CMOS inverter shows a large voltage gain value of 80 and small switching threshold voltage variation of 5% from the ideal value. The static power consumption is estimated to be 0.28 μW (per inverter). The noise margin, estimated from the butterfly plot using maximum equal criterion method, is 3.54 V, which reaches 70.8% of the ideal value. Moreover, the CMOS inverter shows dynamically stable operation using a 100 Hz square waveform input. The output rise and fall delays are measured as 63 and 143 μs, which indicates the logic swing can be used in kHz range using current material and devices. The current result suggests MXene (Ti_3C_2) films have potential as contacts for various electronic applications.

3.5 Experimental Section

Ti$_3$C$_2$ MXene synthesis: 1g of Ti$_3$AlC$_2$ MAX phase powder was slowly immersed into a mixture solution of 20ml 9M HCl and 1g of LiF under magnet stir, which was cooled by icebath to minimize localized heat from initial exothermic reaction. The mixture solution was kept at 35 C with magnet stir in oil bath, for 24 hours. The mixture was transferred to a centrifuge tube and washed several times with additional DI water to make total volume of 50 ml. Each wash was performed by centrifuge at 3000 RCF for 5 min. The supernatant was decanted, and the sediment was redispersed into DI water by hand shaking. Once the pH of the supernatant reached around 6, the final centrifuge was done at 500 RCF for 30 min after redisperse of the sediment in DI water. The supernatant solution containing delaminated 2D MXene flakes was collected for further experiments.

Oxide film growth: Tetrakis(dimethylamido)hafnium (IV) and Diethylzinc (DEZ) were used as the source of Hf and Zn for the deposition of HfO$_2$ (1000 cycles) and ZnO (160 cycles) films by ALD system (Savannah 100, Ultratech) at 160 °C. Deionized water was used as the source of oxygen. The pulse/purge times (in second) for Zn, Hf and O sources used in device fabrication are 0.015/10, 0.02/10 and 0.015/10, respectively. The deposition rate for HfO$_2$ and ZnO layer are 0.14 and 0.15 nm/cycle. The 25 nm p-type SnO film was grown by reactive magnetron sputtering in a mixture gas of Ar and O$_2$, the oxygen partial pressure is 9%.

Device Fabrication: The MXene film was spray-coated on 1-inch glass substrate, followed by growing HfO_2 dielectric. The p-type SnO layer was patterned by lift-off technique and the n-type ZnO film was patterned by wet-etching process using HCl solution. The top MXene contacts were spray-coated and patterned by lift-off technique. The TFTs were annealed at 175 °C for 30 min in tube furnace in ambient.

Material and Device Characterization: The I-V curve of MXene film (1 $inch^2$) was measured by four-point system, using Keithley 2400 source meter, the span between tips is 1 mm. The resistivity *vs* temperature curve was measured in reducing atmosphere (4% H_2 in Ar). Photo-electron spectroscopy in air (PESA) measurement was performed by Riken AC-2 photoelectron spectrometer. X-ray diffraction (XRD) patterns of the films were obtained by a Bruker D8 Advance XRD system. Raman spectra were performed by a LabRAM ARAMIS Raman spectrometer (Horiba Scientific) with a 473 nm laser source excitation. Scanning electron microscopy (SEM) was performed by Nova Nano (FEI). Atomic force microscopy (AFM) was performed by an Asylum Research (MFP-3D) scanning probe microscope in tapping mode. The capacitance-voltage curve for the HfO_2 dielectric was measured by a capacitance meter (Agilent E4981A). The electrical performance of p- and n-type TFTs and CMOS inverters were characterized at room temperature in dark using a semiconductor device analyzer (Agilent B1500A) and a microprobe station (Summit-11600 AP, Cascade Microtech). The 100 Hz square input signals were provided by a waveform generator (Agilent 33220A), where the V_{low} and V_{high} were set as 0 and 10 V, respectively. The output signal was measured from an oscilloscope (Tektronix TDS 2024B).

Chapter 4

Quantum Dot Electrical Double Layer Transistor with MXene Electrical Contacts

4.1 Introduction

Two-dimensional (2D) nanomaterials are essential building blocks in many applications.[102-105] Solution-mediated synthesis of 2D materials is attractive because it allows property tuning by manipulating their sizes, compositions, and surface chemistry. Using dispersions of 2D materials enables cost-effective device fabrication, but significant challenges remain.[106] Such challenges include finding nanomaterial combinations that are compatible with each other and sufficiently stable for conventional electronic fabrication processes (e.g. dry etching and photolithography).

MXenes, a large family of 2D transition metal carbides and/or nitrides, can be prepared by selective chemical etching of ternary layered precursors, and their general chemical formula can be written as $M_{n+1}X_nT_x$ where M is early transition metal, X is carbon and/or nitrogen, T_x represents various surface functional groups (mainly =O, -OH, -F groups), and $n = 1 - 3$.[1] The presence of surface functional groups benefits them with a hydrophilic surface. Aqueous $Ti_3C_2T_x$ MXene dispersion can be directly used to form highly conductive thin films with electrical conductivity up to 10^4 S cm^{-1},[29, 30] which offers significantly simplified fabrication process over carbon-based 2D materials counterparts.

The coexistence of fast charge transport and rich surface chemistry of MXenes featured them in high performance electrochemical energy storage applications,[22, 35] while their electronic/optoelectronic applications are relatively unexplored that we have termed as *MXetronics*.[18, 107] $Ti_3C_2T_x$ has shown its excellent potential as contact materials in oxide TFTs,[30] organic TFTs,[108] and photodetectors,[109] as well as work function modulator in halide perovskite solar cells.[110]

Amongst broad classes of solution-processable semiconductor materials, PbS colloidal quantum dots (CQDs) are promising for optoelectronic applications owing to their size/shape effect on band gap through the quantum confinement effect and high absorption efficiency.[111-113] During the synthesis process, the surface of CQDs is stabilized by long-alkyl chain ligands, namely oleic acid (OA), making them highly dispersible in broad organic solvents. However, such long-ligands should be replaced by short ligands to create electrical paths throughout the CQDs films.[114] The selection of ligands affects the energy level of PbS CQDs,[111] hence their electronic and optical properties can be modulated.[115, 116] Many efforts have been made to functionalize PbS CQD surfaces with short organic ligands (e.g. 3-mercaptoprepionic acid (3-MPA), benzenedithiol, 1,2-ethanedithiol (EDT), thiocyanate, *etc.*) where electron dominant ambipolar characteristics were found in TFT devices.[117-123] In contrast, atomically short halide ligands provide better stability and significantly enhanced *n*-type transport.[124] Few nanometers-sized CQDs result in a high surface area density in their restacked thin film form. This can be a drawback in bottom-gated TFTs with oxide dielectric because it results in significant hysteresis and relatively low carrier mobility due to large carrier trap density.[125] PbS CQDs/dielectric interfacial trapping sites can be significantly reduced in a top-gated structure with hydroxyl-free polymer dielectrics, resulting in electron mobility of 0.2 cm^2 V^{-1} s^{-1}.[118, 122] Furthermore, such porous nanocrystal networks can be effectively gated by ionic liquid/gel,[126] achieving high electron mobility of 1.9 and 2.1 cm^2 V^{-1} s^{-1} for 3-MPA and EDT-treated PbS CQD TFTs, respectively.[117, 127] Despite these significant developments in CQD TFTs, the electrode materials engineering has not been explored beyond the high-energy deposited noble metals.

In this work, we report fully-solution processed electrical double layer transistors (EDLTs) fabricated by using metallic 2D Ti$_3$C$_2$T$_x$ MXene as electrodes, 0D PbS CQD film as semiconductor, and ionic gel consist of 1-hexyl-3-methylimidazolium bis(trifluormethylsulfonyl)imide (HMIM-TFSI) ionic liquid and poly(vinylidene fluoride-co-hexafluoropropylene) (P(VDF-HFP)) gelling polymer as gate dielectric layer. We adopt iodide capping in PbS CQDs to enhance *n*-type transport in a top-gate bottom-contact structure. The relatively low work function of Ti$_3$C$_2$T$_x$ MXene compared to gold and platinum allows favorable electron injection. As a result, our MXene/PbS CQD EDLTs achieve maximum μ_{sat} of 3.32 cm^2 V^{-1} s^{-1}, on/off current ratio of 1.87 × 10^4, low threshold voltage of 0.36 V, and subthreshold swing (SS) of 163 mV dec^{-1} with negligible hysteresis at a small driving voltage range of 1.25 V. This contribution marks the first material integration of MXene and CQDs as well as the first demonstration of dry-etch patterning of MXene electrodes by conventional photolithography process, which opens a wide range of large potential uses of MXenes in future electronic applications.

4.2 Results and Discussion

4.2.1 Device Fabrication and MXene Electrodes Patterning

The fabrication process of the $Ti_3C_2T_x$ MXene/PbS EDLTs is shown in **Figure 4.1**. In brief, aqueous $Ti_3C_2T_x$ MXene solution with a concentration of 1.0-1.5 mg mL^{-1} was spray coated on polyethylene naphthalate (PEN) substrate with thin Al_2O_3 moisture barrier layer. Spray coating is suitable for large-area film deposition **(Figure 4.2(a))** where the amount and concentration of the sprayed solution are directly related to the film thickness and uniformity, respectively. Approximately 300 nm thick MXene films were successfully patterned into the interdigitated source and drain bottom-contacts **(Figure 4.2(b))** by conventional photolithography, followed by the reactive ion etching process and photoresist removal. PbS CQDs were deposited on the top of patterned MXene electrodes using a layer-by-layer (LbL) spin coating process for effective ligand-exchange and good film coverage **(Figure 4.3)**. The device fabrication was completed by spin coating of ionic gel and another MXene layer for top-gate electrodes (See experimental section for detail).

It is worth noting that we choose dry-etch patterning of MXene films instead of the convenient lift-off method used in our previous work,[30] in order to avoid unwanted sidewalls and re-deposition of lift-off residue which are common causes of device failure for top-gate structures with interdigitated electrodes. The successful dry-etch patterning of such porous nanosheet networks on smooth substrates surface requires optimization of photoresist thickness and photolithography parameters; otherwise, MXene films with the cured photoresist layer would undergo irregular detachment from the substrate during the photoresist development process. We believe that this critical issue is not limited to MXene film, but most solution processed 2D nanosheet films as well, due to the lack of strong bonding with the substrate.

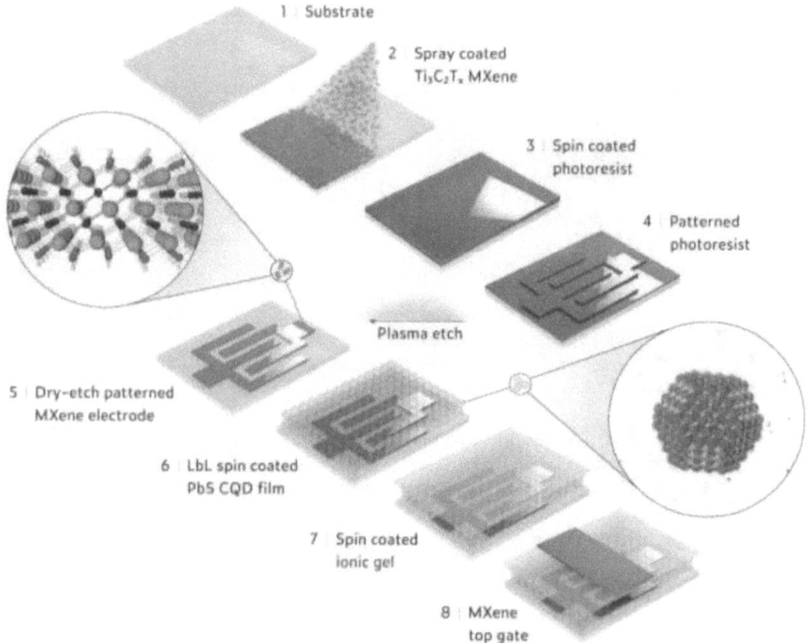

Figure 4.1. Schematic illustration of the device fabrication process flow to make electrical double layer transistors (EDLTs) using n-type iodide-capped PbS CQDs and $Ti_3C_2T_x$ MXene electrical contacts. The MXene electrodes were patterned by conventional photolithography and dry etching process.

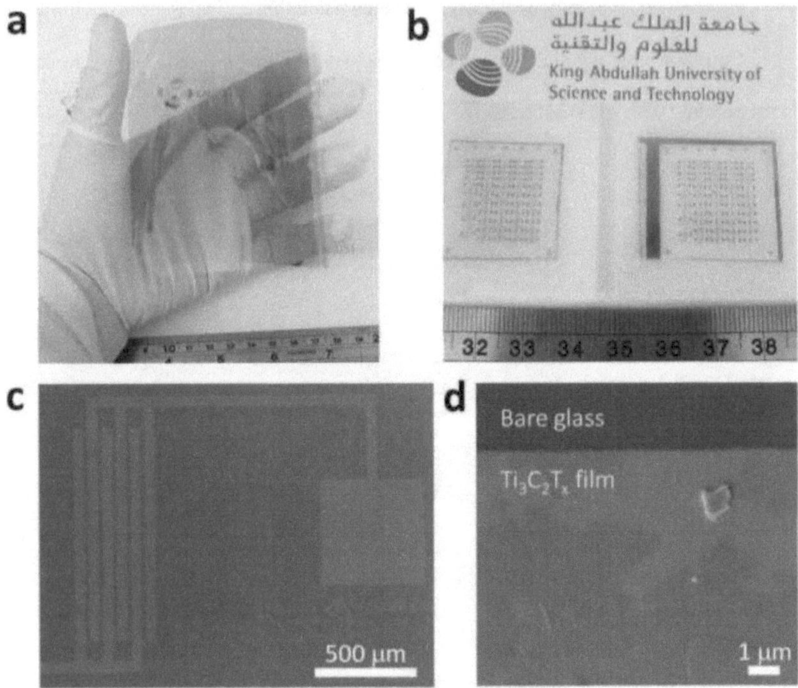

Figure 4.2. a-b) Digital photographs of (a) a spray-coated MXene film on PET substrate, and (b) patterned MXene electrode arrays on PEN substrate. This is the first demonstration of global MXene interconnects patterned by standard positive photolithography and dry etching process. **c-d)** SEM images of patterned $Ti_3C_2T_x$ MXene electrode on glass substrate in (c) low and (d) high magnification. MXene and bare glass are shown in brighter and darker contrast, respectively. The surface of the sample was coated by 3 nm thick iridium layer to avoid charge accumulation.

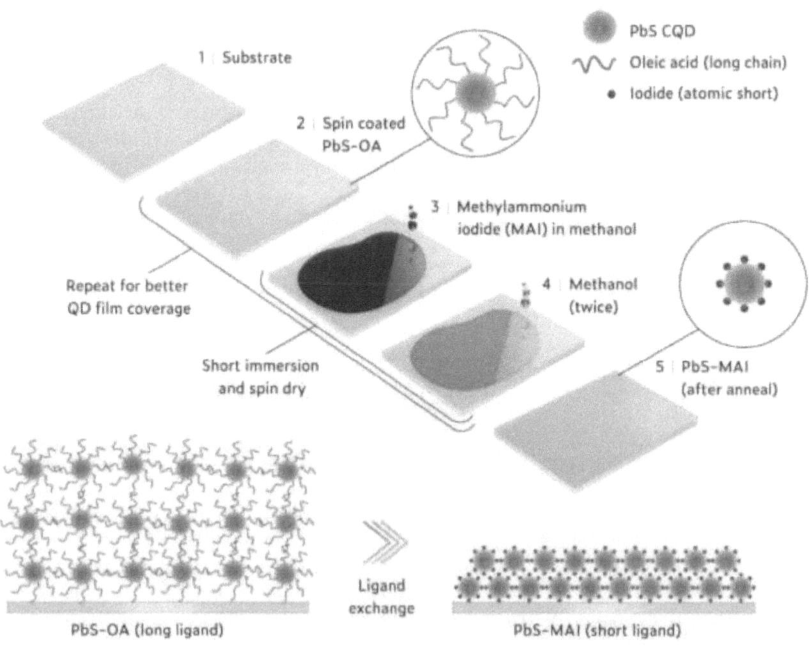

Figure 4.3. (top) Schematic illustration of LbL deposition and ligand exchange process from PbS-OA to PbS-MAI. (bottom) Cross-section illustration of PbS CQD films before and after ligand exchange process.

Figure 4.2(c-d) display scanning electron microscopy (SEM) images of patterned MXene electrodes showing well-defined shapes for the channel width and length of 5000 and 20 µm, respectively, as well as sufficiently good coverage for bulk transport. To the best of our knowledge, this is the first demonstration of successful dry-etch patterning of MXene films by the conventional photolithography process. Moreover, our MXene patterns have much better resolution compared to the reported MXene patterning methods such as laser scribing or direct printing methods.[128, 129] The smallest pattern width is nearly 7 μm as can be seen in the numbers near the contact pad area. Such patterned MXene electrode array is essential for future studies in electronic device applications of MXenes, which are not limited to $Ti_3C_2T_x$ but other members of MXene family with different work function coverage.

4.2.2 Characterization and Work Function Analysis of MXene

Properties of the spray-coated $Ti_3C_2T_x$ MXene films are shown in **Figure 4.4**. The X-ray diffraction (XRD) patterns of $Ti_3C_2T_x$ MXene films with different thicknesses are shown in **Figure 4.4(a)** along with their optical images. All observed peaks belong to basal-plane diffraction, suggesting successful exfoliation and highly c-oriented restacking nature of MXene nanosheets. The first (0002) peak becomes stronger for thicker films along with the evolution of its multiple order diffraction peaks, suggesting long-range ordered stacking. The calculated c-lattice parameter (c-LP) is 25.8 Å (interlayer space of 12.9 Å) based on the (0002) peak position of ~ 6.84°, suggesting the presence of water molecules associated with Li^+ cations between MXene nanosheets.[130] It has been reported that strongly-confined water molecules can enhance EDL capacitance due to their negative dielectric constant,[131] and the negative capacitance has been applied to MoS_2 transistor devices, achieving ultralow subthreshold swing of 6.07 mV dec^{-1}.[132] The characteristic Raman spectra of MXene films (**Figure 4.4(b)**) show no thickness dependence. A small variations of the peak shape at 725 cm^{-1} are observed which may be affected by slightly different local populations of oxygen and hydroxyl surface functional groups. The observed peak center belongs to the out-of-plane vibration of central Ti-C bonds for oxygen surface functional groups, while the same vibration mode for hydroxyl terminated MXene is 708 cm[-1].[33] It is worth to note that the presence of this peak can be a signature of large sized MXene flakes.[133] **Figure 5.4(c)** shows the potential of MXene films as transparent, flexible, conducting electrodes compared to silver nanowire (AgNW),[134] PEDOT:PSS,[135] SWCNT,[136, 137] and graphene.[96] The optical figure of merit of our MXene films is 12.6 (**Figure 4.5**), slightly lower than the record value of 15.[29] This is likely because the spray-coated MXene films are not vacuum annealed prior to the measurements in ambient air, and the intercalated water molecules result in larger sheet resistance and possible light scattering.

To evaluate the potential of MXenes in flexible applications, we monitored the two-terminal resistance of MXene films deposited on poly(ethylene terephthalate) (PET) substrate in Figure 1(b) with an average sheet resistance of $119 \pm 5 \; \Omega \; sq^{-1}$. As shown in **Figure 4.4(d)**, MXene films show excellent electromechanical stability, where the resistance increases by 1 % for a bending radius of 2.0 mm, and by 2.8 % for an extreme bending radius of 0.2 mm. The resistance value is restored to near the initial value once the strain is removed. This is due to the restacking nature of 2D nanosheets where new percolation paths are created under strain. Once the film undergoes 1000 cycle testing from a radius of 1.8 cm to 9.6 cm, the film resistance increases by 6.5 %. (**Figure 4.6**). The incremental accumulation of irreversible changes suggests a decrease in the number of percolation pathways and an increase in their porosity due to repeated strain. Proper encapsulation may be required to maintain their electromechanical performance for flexible applications.

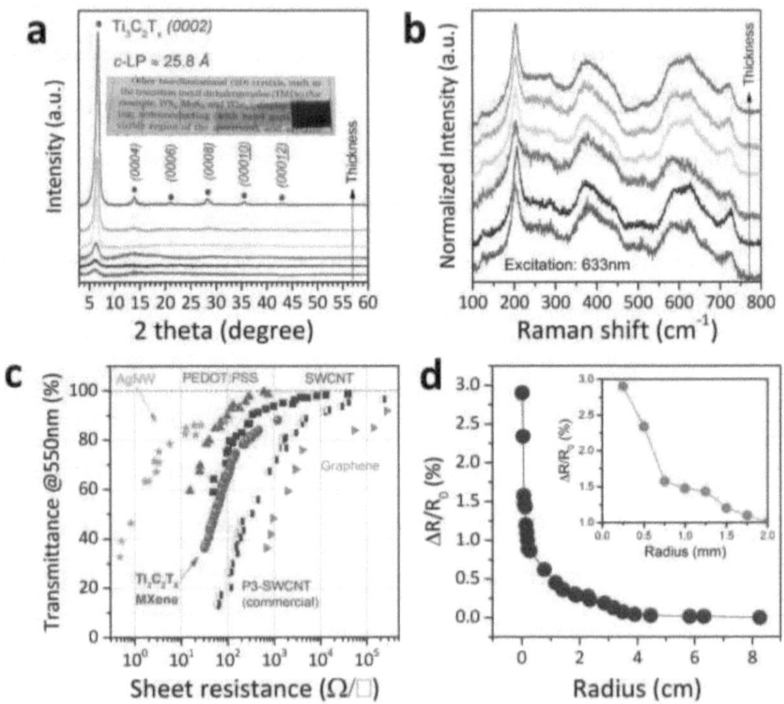

Figure 4.4. a) Out-of-plane XRD patterns, and **b)** Raman spectra of Ti₃C₂Tₓ MXene films on glass substrates. Inset: Photograph of MXene films with different thicknesses. **c)** Transmittance of MXene films as a function of sheet resistance. Data for AgNW and other non-MXene materials are adopted from ref.[134] and ref.[29], respectively. **d)** Two-terminal resistance variation as a function of bending radius. Initial resistance was approximately 600 Ohm.

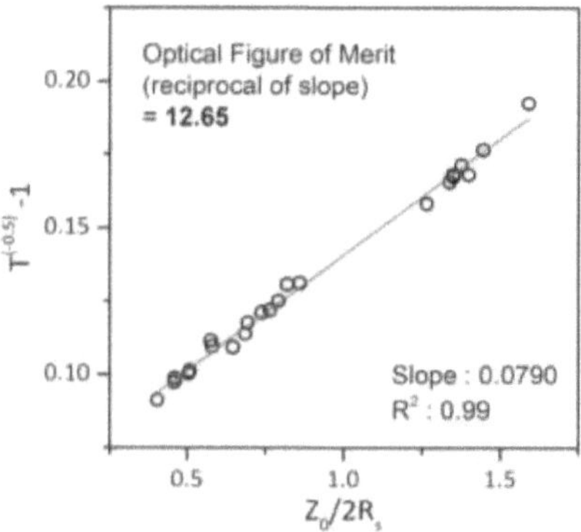

Figure 4.5. Fitting curve for the optical figure of merit of $Ti_3C_2T_x$ MXene films, based on a relationship: $T^{-0.5} - 1 = \frac{\sigma_{OP}}{\sigma_{DC}} \frac{Z_0}{2R_s}$, where T is transparency at 550 nm; σ_{DC} and σ_{OP} are DC and optical conductivities, respectively; Z_0 is the free space impedance (377 Ohm); R_s is sheet resistance.

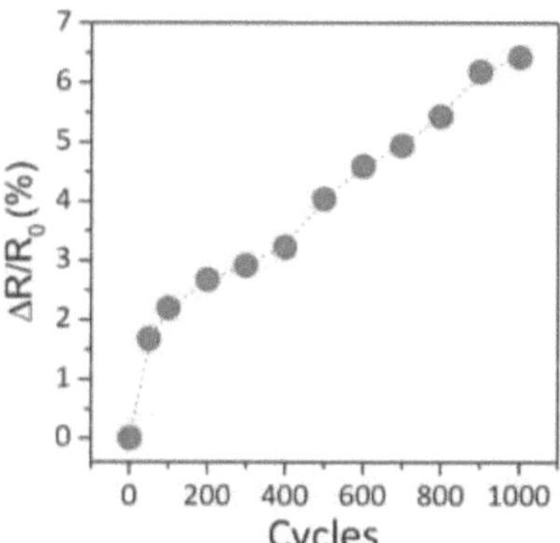

Figure 4.6. Variation of two-terminal resistance of $Ti_3C_2T_x$ MXene film on PET substrate, under 1000 cycle of convex bending tests at radius of 1.8 cm to 9.6 cm.

We carefully monitored the work function of $Ti_3C_2T_x$ MXene films (Φ_{MXene}) by photoelectron spectroscopy in air (PESA) and ultraviolet photoelectron spectroscopy (UPS) techniques as shown in **Figure 4.7**. We found that Φ_{MXene} is significantly affected by the ambient conditions of MXene films storage, suggesting possible charge interactions with surrounding molecules/moisture similar to some reported 2D materials.[138, 139] Using the PESA method, we could confirm a large window of Φ_{MXene} ranging from 4.44 to 5.15 eV as shown in **Figure 4.7(a).** The Φ_{MXene} values of our as-prepared MXene films using fresh solution is 4.7 to 4.9 eV, where slightly larger values were found when the number of washing process was increased by repeating centrifuge and re-dispersion in DI water (**Figure 4.8**). This suggests that Φ_{MXene} is strongly correlated to the synthesis protocols and surface functional groups. Φ_{MXene} increases by two time-dependent factors: (1) exposure to ambient air after the formation of MXene thin film and (2) storage of aqueous MXene solution prior to film deposition (**Figure 4.9**). For MXene thin film deposited by fresh suspension, Φ_{MXene} increases from 4.90 eV to 5.10 eV after exposure to ambient air and saturates within a day. In the case of MXene suspension stored at 3 °C for 24 hours before film deposition, slightly higher Φ_{MXene} is found (5.02 eV to 5.15 eV). This can be attributed to surface reaction involving dissolved oxygen despite the low storage temperature and relatively short storage time,[140] and possible changes in the surface functional group population upon the change of environment. In contrast, Φ_{MXene} decreases over time when the film is stored in nitrogen-filled glovebox with minimal moisture.

However, such environmental changes of Φ_{MXene} were not observed by UPS in ultrahigh vacuum (**Figure 4.7(b)**), Φ_{MXene} measured by UPS is 4.42 and 4.47 eV for N_2-stored and air-stored MXene films, respectively. Moreover, their X-ray photoelectron spectroscopy (XPS) spectra do not show significant differences or any indication of oxidation to TiO_2, regardless of their storage ambient (**Figure 4.10**). Hence the modulation of Φ_{MXene} is most likely due to the interaction with intercalated/surrounding moisture and gas molecules, associated with reversible changes in the oxidation states of Ti atoms by mean of charge compensation.

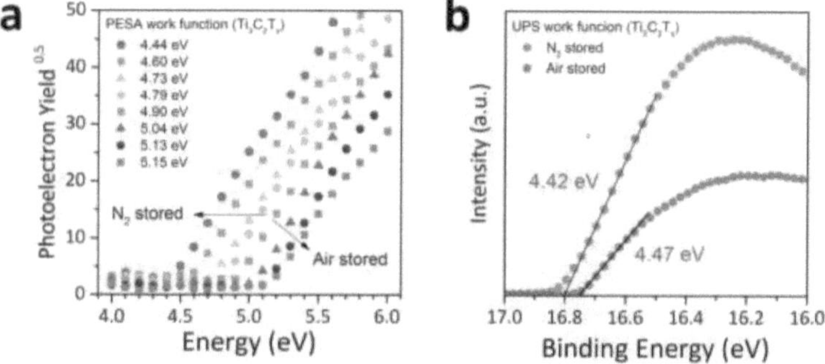

Figure 4.7. a) PESA measurement of $Ti_3C_2T_x$ MXene films. The work function range of 4.44 to 5.15 eV was experimentally confirmed depending on the film storage condition. **b)** UPS Helium $I\alpha$ ($h\nu = 21.21$ eV) spectra under ultrahigh vacuum for secondary electron cutoff region of MXene films.

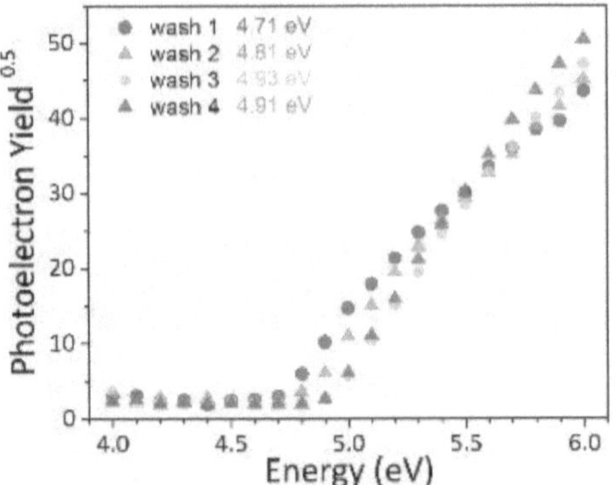

Figure 4.8. PESA spectra of Ti$_3$C$_2$T$_x$ MXene films made of collected suspension solutions at different number of wash cycles (1–4) after Li$^+$ intercalation. Each MXene suspension was collected during the repeated wash cycles, and corresponding films are sprayed coated immediately and measured by PESA in a short time.

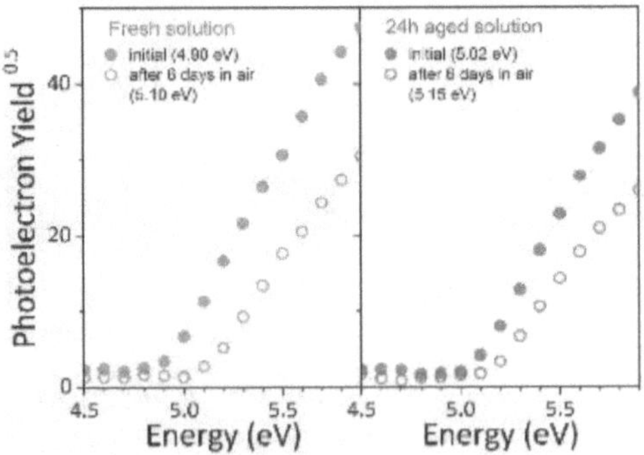

Figure 4.9. PESA spectra of Ti$_3$C$_2$T$_x$ MXene films made of different MXene solution storage time; fresh solution (red, left) and 24 hours aged solution in refrigerator at 3 °C (blue, right). Filled and opened symbols represent the pristine and air-stored sample, respectively. The exposure of spray-coated MXene film into air and the storage of MXene suspension in refrigerator prior to spray deposition result in the increase of work function.

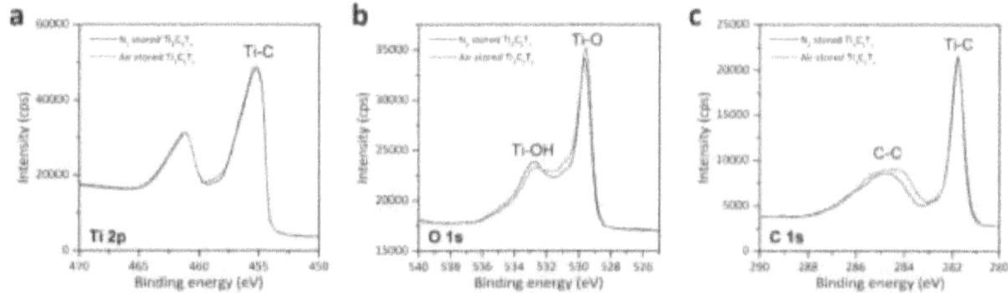

Figure 4.10. XPS spectra of Ti₃C₂Tₓ MXene films stored in dry N₂ and ambient air. a) Ti 2p, **b)** O 1s, **c)** C 1s core levels. No evidence of significant oxidation, such as evolution of Ti^{4+} peak or weakening in Ti-C bonding, was found.

4.2.3 Optical and Morphological Characterization of PbS CQDs

Figure 4.11(a) shows the Fourier transform infrared (FTIR) spectra of PbS CQD films showing effective ligand exchange (LE) from organic OA to atomic iodide capping by the elimination of C-H and COO bond stretching modes at around 2900 and 1500 cm^{-1}, respectively. Enhancement in inter-dot coupling is evidenced by red-shift in the first excitonic peaks in UV-vis absorption spectra (**Figure 4.11(b)**) and photoluminescence peaks (**Figure 4.11(c)**).[114] To identify the highest occupied molecular orbital (HOMO) of the iodide-capped PbS CQD films, we employed PESA method as shown in **Figure 4.11(d)**. From the observed HOMO level at 5.37 eV and an optical band gap of 1.23 eV, we can deduce the lowest unoccupied molecular orbital (LUMO) of iodide-capped PbS CQDs to be approximately 4.14 eV. The calculated LUMO level is in good agreement with literature for iodide-capped PbS CQDs,[115, 141, 142] and is a suitable value for n-type transport using Ti₃C₂Tₓ MXene contacts.

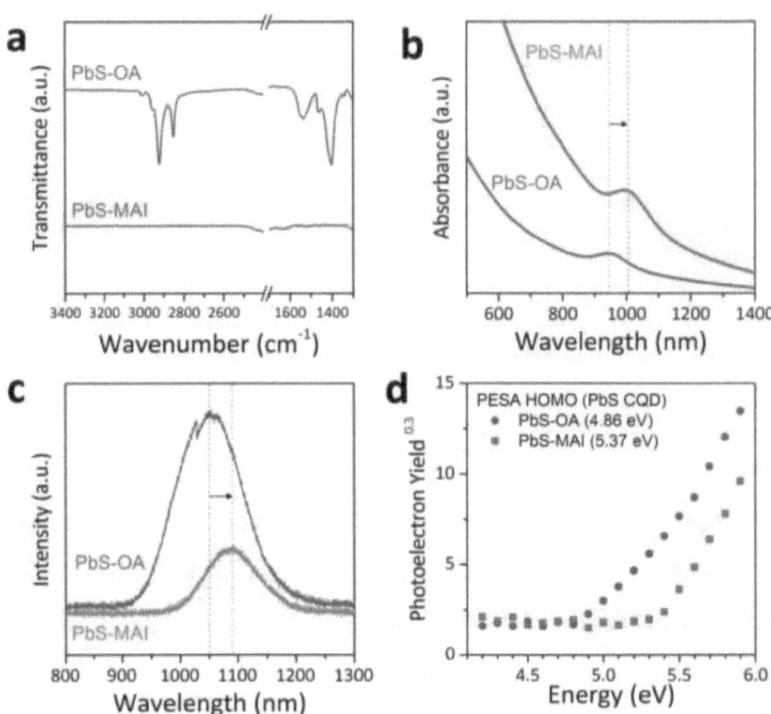

Figure 4.11. a) Fourier transform infrared (FTIR) spectra, **b)** UV-vis-NIR absorption spectra, **c)** Photoluminescence curves of PbS CQD films with OA ligand (PbS-OA) and iodide-capping (PbS-MAI). **d)** PESA measurement of PbS-OA and PbS-MAI CQD films.

Transmission electron microscopy (TEM) images of PbS CQD before and **after** ligand exchange are shown in **Figure 4.12(a-b)**. It can be clearly seen that the OA-capped PbS CQDs are well isolated from each other, while the iodide-capped PbS CQDs have shorter inter-dot spacing. Such dramatic decrease in inter-dot spacing may cause incomplete coverage of the CQD films, so we repeated the deposition and ligand exchange process using LbL approach to improve the coverage. The surface morphology of PbS CQD films was evaluated by atomic force microscopy (AFM) and SEM as shown in **Figure 4.12(c-d)**, which confirm the presence of percolation networks despite the presence of micro-cracks.

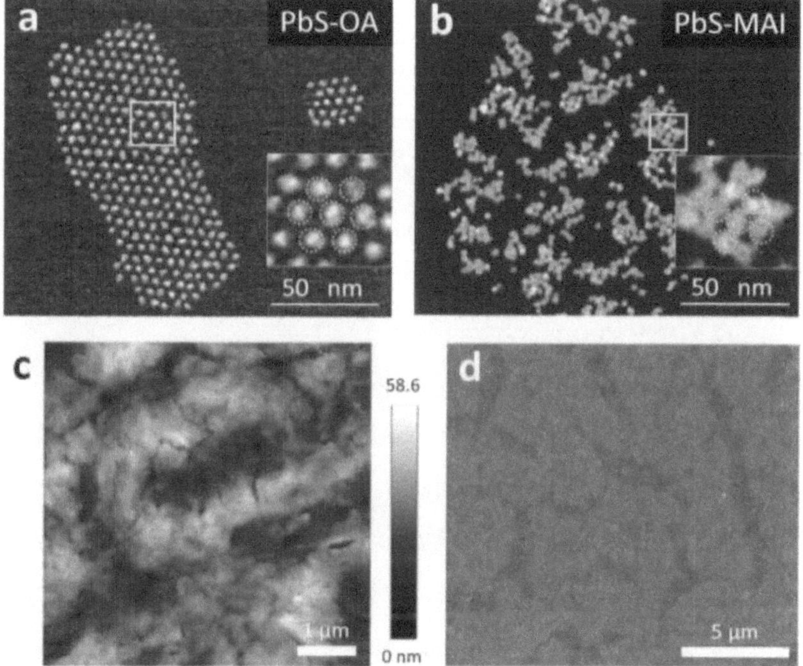

Figure 4.12. a-b) TEM images of the CQD films (a) before and (b) after the ligand exchange process. Inset is the enlarged views of the area in the yellow-colored boxes. PbS-OA has relatively long ligand that is electrically insulating. **c)** AFM image, and **d)** SEM image of the iodide-capped PbS CQD film showing the presence of percolation network for electrical conduction.

4.2.4 $Ti_3C_2T_x$ MXene/PbS EDLT Device Performance

The output curve of the MXene/PbS EDLT device in **Figure 4.13(a)** shows good n-type transport behavior and successful suppression of the ambipolar secondary-carrier transport at high V_{DS}, which was found in 3-MPA capped PbS CQDs.[117] This result can be attributed to two factors: (1) monovalent iodide capping by means of n-doping, and (2) low Φ_{MXene} that allows effective electron injection while blocking hole injection. The majority carriers of iodide-capped PbS CQD films are electrons as evidenced by both the negative Hall and Seebeck coefficients.[143] Similarly, strong n-type transport behavior was achieved for iodide-capped PbS CQD TFTs using silver electrodes with relatively low work function.[144] The drain current increases linearly at low V_{DS} (< 0.25 V) and saturates at high V_{DS} (> 0.5V). **Figure 4.13(b)** shows the transfer curve of the optimal device at V_{DS} = 0.5 V, showing a typical enhancement mode characteristics (channel width, W = 6000 μm, channel length, L = 100 μm). The turn on voltage (V_{on}) is 0.02 and 0.10 V for forward and reverse sweep, respectively. The threshold voltage (V_{th}) is 0.36 V for both sweeps, indicating negligible hysteresis. The subthreshold region is well observed between V_{on} and V_{th} in the semi-logarithmic scale curve, where the subthreshold swing (SS) value is 167 and 181 mV dec^{-1} for forward and reverse sweep, respectively. The on/off current ratio (I_{on}/I_{off}) is 6.2×10^3. The maximum gate leakage current (I_{GS}) remained at 2×10^{-7} A (**Figure 4.14**), which is a typical value for ionic-gel gated devices.[145]

Figure 4.13(c) shows the effective areal capacitance of the ionic gel as a function of frequency (50 mHz to 100 kHz) obtained from electrochemical impedance spectroscopy

(EIS) analysis by an equation of $C = (2\pi f\, Z'')^{-1}$, where C is capacitance, f is frequency, and Z" is imaginary impedance. The ionic gel exhibits two distinct frequency regions where the capacitance changes: (1) from high frequency to 35 Hz (semicircle part of the Nyquist plot), where the alignment of dipoles in P(VDF-HFP) polymer contributes to the capacitance, and (2) from 35 Hz to low frequency (diagonal line with a slope of 45° in the Nyquist plot), where the diffusion of ionic liquid ions starts the EDL formation. The capacitance reaches to 7.3×10^{-7} F cm^{-2} at 177 mHz which is the point of fully completed EDL formation. The ionic gel shows a considerably large charge-transfer resistance of ~ 2.4 MΩ which evidences the absence of major pseudocapacitive reaction between the electrode and the ionic gel electrolyte but the existence of a weak electrochemical behavior. Assuming the EDL formation was completed during the transistor measurement, the estimated μ_{sat} is 1.99 cm^2 V^{-1} s^{-1} from the equation of $\mu_{sat} = \dfrac{2L}{W}\dfrac{1}{C}\left(\dfrac{\partial I_{DS}^{1/2}}{\partial V_{GS}}\right)^2$.

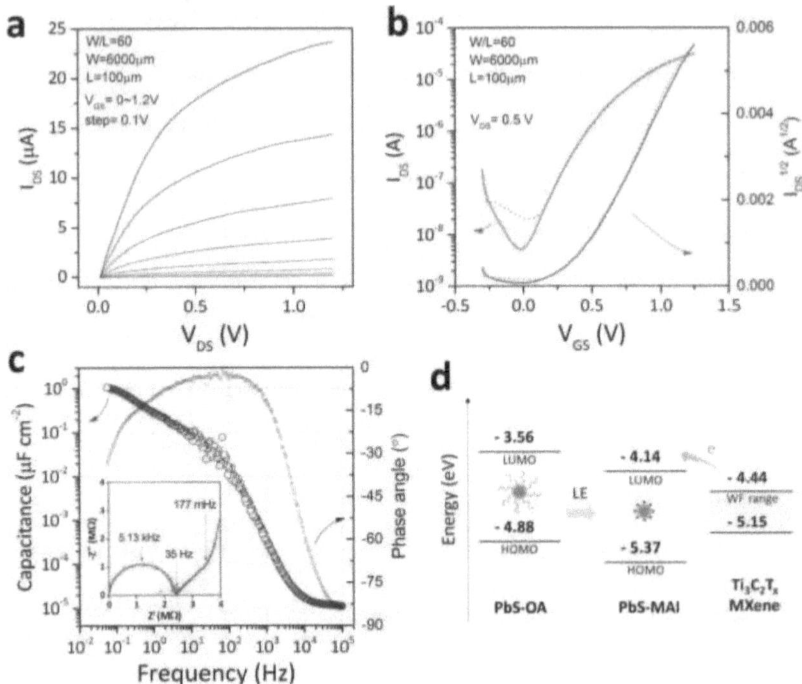

Figure 4.13. a) Output characteristics, and **b)** Transfer characteristics of $Ti_3C_2T_x$ MXene/PbS CQD EDLTs. Solid and dashed lines represent forward and reverse sweep, respectively. **c)** Effective areal capacitance of the ionic gel and phase angle as a function of frequency. Inset is the corresponding Nyquist plot. **d)** Schematic energy band diagram of PbS-OA, PbS-MAI, and $Ti_3C_2T_x$ MXene.

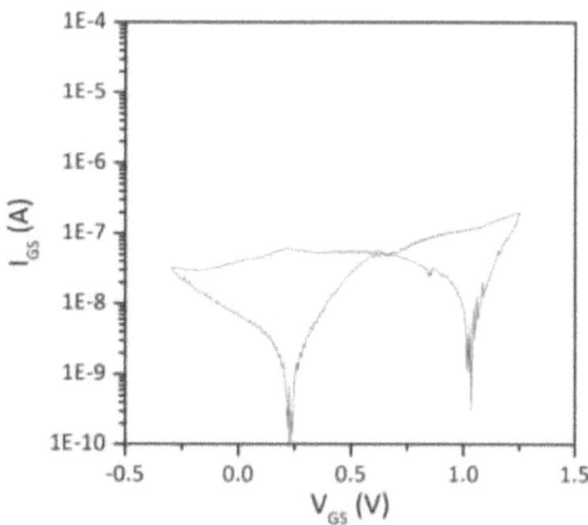

Figure 4.14. Typical gate leakage current of MXene/PbS CQD EDLT.

Many devices were measured to evaluate the variation in the hysteresis behavior. We find that the hysteresis behavior shows variations at low V_{DS} (0.2 V), but it is well suppressed at higher V_{DS} (0.5 V). (**Figure 4.15(a)**). The μ_{sat} and I_{on}/I_{off} can reach 3.32 cm^2 V^{-1} s^{-1} and 1.87×10^4, respectively, for the device with ΔV_{th} = - 0.1 V. (W = 5000 μm, and L = 40 μm) Although the origin of the hysteresis is unclear at this moment, it is likely due to the incomplete penetration of the ionic gel through the small pores in CQD films and/or the local variation of inter-dot coupling. Our observation partially agrees with a recent study on ionic liquid-gated WS$_2$ transistors, where the hysteresis is found to be inversely proportional to lateral flake size.[146] One may notice potential contributions from the electrochemical interaction between MXene and ionic gel considering their excellent performance in supercapacitors. Well-stacked and dried MXene film is, however, electrochemically inactive in ionic liquids, while pre-intercalation of ionic liquid into MXene film can boost their electrochemical capacitance more than 70 times.[147]

We believe that the low V_{th} achieved in this work (0.36 V for MXene/PbS CQD vs. 0.75 V for Au/PbS CQD, **Figure 4.15(b)**) is due to two possible factors. First, the lower work function of $Ti_3C_2T_x$ MXene can allow formation of a better band alignment at MXene/PbS CQD interface (**Figure 4.13(d)**) compared to the large work function of Au, which results in a lower barrier height for electron transport. The reduced value of V_{th} seems to match well with the work function difference, however we cannot completely rule out the possible Fermi level pinning effect at Au/CQD interface.[148] In such a case, the Schottky-Mott rule does not apply, and the barrier height becomes less sensitive to the work function of contact metal. Second, the strong negative surface charge of $Ti_3C_2T_x$ MXene can play an important role by attracting positive cations,[149, 150] which can help cations accumulation at lower gate bias. The charge-induced cation accumulation can promote electron injection from the MXene electrode to PbS CQD channel, which slightly increases I_{off} but reduces V_{th}. Thus the family of MXenes can be alternative electrode materials when low work function metals such as Al and Ag are not compatible with the designed electrolyte. Moreover, the hydrophilic surface of MXene may facilitate the formation of solution-processed films with better coverage at the interface.

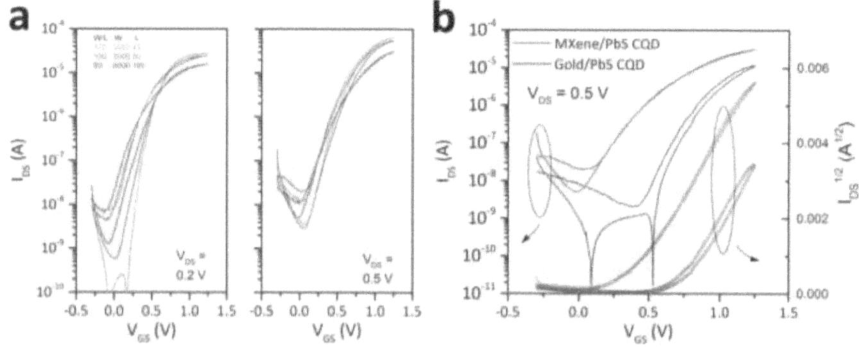

Figure 4.15. a) Transfer curves of MXene/PbS CQD EDLTs at low V_{DS} (0.2V, left panel) and high V_{DS} (0.5V, right panel). The hysteresis is clearly suppressed at higher V_{DS}. Slightly different hysteresis was found from different devices. In general, the electron mobility is proportional to the degree of hysteresis. **b)** Transfer curves of PbS CQD EDLTs with different electrode material; $Ti_3C_2T_x$ MXene (red) and gold (blue). The threshold voltage of MXene and gold contacted devices are 0.36 V and 0.75 V, respectively.

4.3 Conclusions

We have successfully demonstrated electric double layer transistors by combining 2D $Ti_3C_2T_x$ MXene, 0D PbS CQDs, and ionic gel using all solution process. $Ti_3C_2T_x$ MXene films have been patterned by conventional photolithography and reactive ion dry etch process for the first time, thus demonstrating their potential in large scale electronic applications. The relatively low work function of highly conductive $Ti_3C_2T_x$ MXene (4.4 eV in ultrahigh vacuum) is suitable for n-type transport of iodide-capped PbS CQD with their LUMO level of 4.14 eV. The small band offset at MXene/CQD interface allowed our MXene/PbS EDLT achieving a high electron saturation mobility of up to 3.32 cm^2 V^{-1} s^{-1}, switching ratio of 1.87 × 10^4, threshold voltage of 0.36 V, subthreshold swing of 167 mV dec^{-1} at a low driving voltage of 1.25 V with negligible hysteresis. The negative surface charge of MXene contributes to the accumulation of cations at lower gate bias. Our results show a large potential of MXenes as contact materials for electronic devices, which can be further generalized to other members of MXene family with a wider work function coverage, beyond $Ti_3C_2T_x$.

4.4 Experimental Section

PbS CQD synthesis: CQD synthesis was performed following previously reported methods.[151] In a typical process, 0.18 gram (1 mol) of bis(trimethylsilyl)sulfide (TMS) was added to 10 ml of 1-octadecene (ODE) that had been dried and degassed by heating to 80 °C under vacuum for 24 hours. A mixture of OA (1.34 g, 4.8 mmol), PbO (0.45 g, 2.0 mmol) and ODE (14.2 g, 56.2 mmol) was heated to 95 °C under vacuum for 16 hours and placed under Ar gas. The flask temperature was increased to 120 °C and the TMS/ODE mixture was injected. After injection, the flask was allowed to cool gradually to 35 °C. The CQDs were precipitated using distilled acetone (50 ml) and centrifuged. The supernatant was discarded, and the precipitate was re-dispersed in toluene. The CQDs were precipitated again using methanol (20 ml), centrifuged (5 min), dried, dispersed in octane (50 mg ml^{-1}) and transported into a nitrogen glovebox (oxygen below 2 ppm and moisture below 10 ppm).

Ti_3C_2 MXene synthesis: The synthesis of $Ti_3C_2T_x$ MXene was performed by previously reported methods with minor modification.[152] 1g of Ti_3AlC_2 MAX phase powder was slowly immersed into a mixture solution of 6 ml HCl (12 M), 1 ml HF (49 %), and 3 ml DI water, under magnet stir, which was cooled by icebath to minimize localized heat from the initial exothermic reaction. The mixture solution was kept at 40 °C with a magnet stir in an oil bath, for 15 hours. The mixture was transferred to a centrifuge tube and washed several times with additional DI water to make a total volume of 50 ml. Each wash was performed by centrifuge at 3000 RCF for 5 min; the supernatant was decanted, and the sediment was redispersed into DI water by hand-shaking. Once the pH of the supernatant reached around 6, the sediment was redispersed into aqueous LiCl solution (0.75 M) and stirred for 20 minutes. Li^+ intercalated multilayer MXene sediment was washed by DI water for additional 2 centrifuge cycles to avoid salt in the solution, regardless the high concentration of exfoliated MXene at this point. The final centrifuge was done at 500 RCF for 10 mi, and the supernatant solution containing delaminated 2D MXene nanosheets were collected for next three cycles and used for spray coating.

Ionic gel synthesis: 80 mg of poly(vinylidene fluoride-co-hexafluoropropylene) (P(VDF-HFP)) pellets were dissolved in 1 ml of acetonitrile. Then the solution was stirred at 60 °C overnight. 20 mg of 1-hexyl-3-methylimidazolium bis(trifluormethylsulfonyl)imide (HMIM-TFSI) ionic liquid was added into the P(VDF-HFP) solution and stirred for 3 hours at room temperature.

Device fabrication: Aqueous $Ti_3C_2T_x$ MXene suspension with a concentration of $1.0 - 1.5$ mg ml^{-1} was spray coated on clean substrates. Patterning of MXene film was achieved by conventional photolithography process. AZ1512HS photoresist (MicroChemicals) was spin coated at 3000 rpm for 30 s, baked at 100 °C for 1 min, exposed to UV light (40 mJ cm^{-2}), and developed by AZ 726 MIF developer (MicroChemicals) for 20s. The plasma etch was performed by using mixture gas of Cl_2/Ar for up to 20 s with ICP power of 1500 W. After patterning, the remaining photoresist was removed by ultrasonication in acetone. The patterned MXene electrodes were transferred to N_2-filled glovebox for PbS CQD deposition by a layer-by-layer spin coating method. For each layer, 20 mg ml^{-1} solution of PbS in hexane was spin-coated at 1000 rpm for 30 s, to form a single monolayer film of the QDs layer. Subsequently, the PbS film was soaked in MAI solution in methanol for 20 s to replace the insulating oleic acid, and then spin-coating at the same speed to remove the excess ligand exchange solution. After that, pure methanol was dropped on the film to remove unbound MAI and native oleic acid ligands, and this step was repeated twice to ensure all the excess ligands has been removed. PbS layer deposition and ligand exchange were repeated twice in this work. After deposition, the devices were annealed at 120 °C for 30 min to remove residual solvent and to promote coupling between CQDs which can help to improve the conductivity of the channel layer without sintering the CQDs. The ion gel solution was then spin-coated on the PbS CQD film and annealed at 100 °C for 2 hours. Finally, MXene gate was deposited on the top of the channel area.

Material Characterization: X-ray diffraction (XRD) patterns of the films were obtained by a Bruker D8 Advance XRD system using Cu Kαradiation. Raman characterization was performed by a LabRAM ARAMIS Raman spectrometer (Horiba Scientific) with a 633 nm laser source excitation. The thickness and surface morphology of PbS QDs and MXene films were determined by AFM (Bruker Dimension ICON). The morphology images were acquired using the FEI Nova Nano 630 SEM. Photo-electron spectroscopy in air (PESA) measurement was performed by Riken AC-2 photoelectron spectrometer. The electrical performance of MXene film was characterized at room temperature using a Keithley 2400 SourceMeter. The device characterization was done by Keysight B2912A precision/measure unit connected to a probe station in a glovebox. The UV-Vis-Infrared absorbance spectra were measured using Cary 5000 ultraviolet-visible (UV-Vis) spectrometer (Agilent Technologies). The Fourier-transform infrared spectroscopy (FTIR) measurements were recorded on films deposited on glass substrates using the Cary 680 with attenuated total reflectance (ATR) mode in air. The TEM sample was observed by Titan Cs Probe STEM operating at an accelerated voltage of 200 kV. The CQDs were deposited on the carbon side of a holey carbon-copper TEM grid by drop-casting from a 2 mg/ml toluene solution, followed by immersion in the ligand exchange solution and solvent, and then the samples were dried at 80 °C in vacuum. Electrochemical impedance spectroscopy was tested by VMP-3 (Biologic). XPS was carried out in an Axis Ultra DLD spectrometer (Kratos Analytical, UK) equipped with a monochromatic Al Kα X-ray source (hν = 1486.6 eV) operating at 150 W, a multichannel plate, and delay line detector under a vacuum of ~10^{-9} mbar. Binding energies were referenced to the C 1s binding energy of adventitious carbon contamination, which was taken to be 284.8 eV.

Chapter 5

Thermoelectric Properties of Mo-based MXenes

5.1 Introduction

A large class of two-dimensional (2D) transition metal carbides and nitrides, also known as MXenes, has been under extensive investigation since its first discovery in 2011.[2] These atomically thin materials, with general formula of $M_{n+1}X_nT_x$, are usually obtained by selective removal of the 'A-element' (Al, Ga, etc.) layers from their ternary layered $M_{n+1}AX_n$ phase and similar layered precursors. In MXenes, M is an early transition metal, X is carbon and/or nitrogen, T_x represents the surface functional groups such as OH, O, and/or F groups, and n = 1, 2, or 3. Twenty different MXenes[1, 10] have been experimentally synthesized for applications ranging from energy storage,[3, 22, 27, 153-156] to electromagnetic interference shielding,[14] bio-sensing,[78, 157, 158] ion sieving,[149] water purification,[159] and anti-bacterial activity.[160] As there are more than 70 different MAX phases that have been experimentally prepared, many more MXenes are expected to be explored.[161]

Theoretical studies have predicted that most MXenes are metallic, but few of them (Sc_2CT_x, Ti_2CT_x, Zr_2CT_x, and Hf_2CT_x) are predicted to have a non-zero band gap when terminated with appropriate surface groups.[7, 48, 53] These semiconducting MXenes are expected to have a high Seebeck coefficient at low temperatures; e.g. ~1140 $\mu V\ K^{-1}$ and ~2200 $\mu V\ K^{-1}$ at 100 K predicted for Ti_2CO_2 and $Sc_2C(OH)_2$, respectively. According to a later calculation study, Mo_2CF_2 (with fluorine-terminated surface) has been reported to have superior thermoelectric power factor among the other 35 different functionalized M_2XT_x MXene systems.[57] However, those theoretical studies have assumed a fully controlled surface termination, which may be challenging to achieve in experimental studies, where multiple functional groups co-exist. Hope *et al* reported that the portion of the surface termination groups of $Ti_3C_2T_x$ MXene is dependent on the synthesis routes, and experimentally quantified by ^1H and ^{19}F nuclear magnetic resonance spectroscopy,[37] where HF etching route results in higher proportion of fluorine termination.

In this work, we report the thermoelectric properties of three kinds of Mo-based MXenes - M_2CT_x, $M_3C_2T_x$, and $M_4C_3T_x$. Freestanding MXene papers were fabricated by vacuum-assisted filtration of suspensions of delaminated 2D Mo_2CT_x, $Mo_2TiC_2T_x$, and $Mo_2Ti_2C_3T_x$. The latter two compositions were selected instead of $Mo_3C_2T_x$ and $Mo_4C_3T_x$, which are not thermodynamically stable and thus difficult to synthesize.[26, 162] $Mo_2TiC_2T_x$, and $Mo_2Ti_2C_3T_x$ belong to the family of ordered double-transition-metal carbide MXenes,[38] in which Mo atoms are preferentially located in the outer layers, sandwiching the layers of Ti. As a result, all three Mo-based MXenes that we synthesized in this study have Mo atoms on the surface, creating a similar surface chemistry. Additionally, these Mo-based MXenes with oxygen termination have been reported as 2D topological insulators, based on calculation along with the consideration of spin-orbit coupling.[50-52] 2D topological insulators with controlled size have been predicted to show anomalous Seebeck effects,[163] and the best thermoelectric material, Bi_2Te_3, and its alloys are also topological insulators.[164] This provided additional motivation for studying transport properties of these MXenes.

5.2 Materials Characterization

Figure 5.1 shows a schematic illustration of the MXene synthesis process. The ternary and quaternary carbide precursor powders (Mo_2Ga_2C,[165] Mo_2TiAlC_2,[166] and $Mo_2Ti_2AlC_3$[25]) were first chemically etched by hydrofluoric acid (HF) which attacks the metallic M-A bond.[1] After washing in deionized water several times, these multilayer MXenes (HF-etched powders) were further reacted with tetrabutylammonium hydroxide (TBAOH) which leads to exfoliation of 2D MXene suspensions by intercalation of tetrabutylammonium cation (TBA^+) between the negatively charged MXene sheets, followed by further washing and sonication. Details of the synthesis process are explained in the experimental section. Scanning electron microscopy (SEM) images of the precursors and multilayer MXenes (**Figure 5.2**) show the typical layered structure of the ternary and quaternary carbides and separations between layers after chemical etching in the multilayer MXenes.

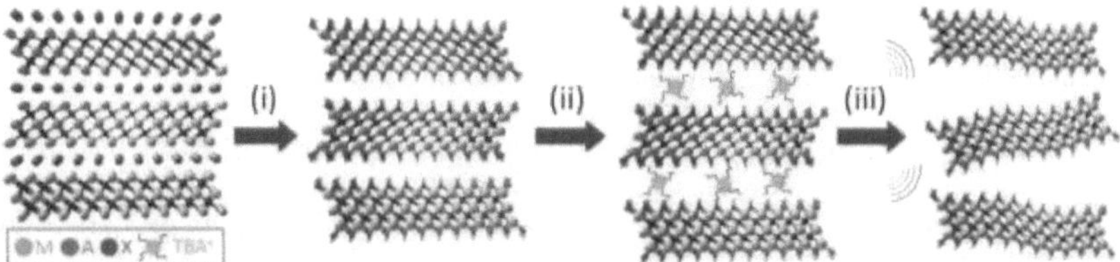

Figure 5.1. Schematic illustration of MXene synthesis process. (i) selective etching of the A-element in the layered precursor materials, (ii) intercalation of organic molecules, tetrabutylammonium cations (TBA$^+$), (iii) exfoliation of 2D MXene flakes by molecule exchange with water and ultra-sonication. Reprinted with permission.[19] Copyright 2017 American Chemical Society.

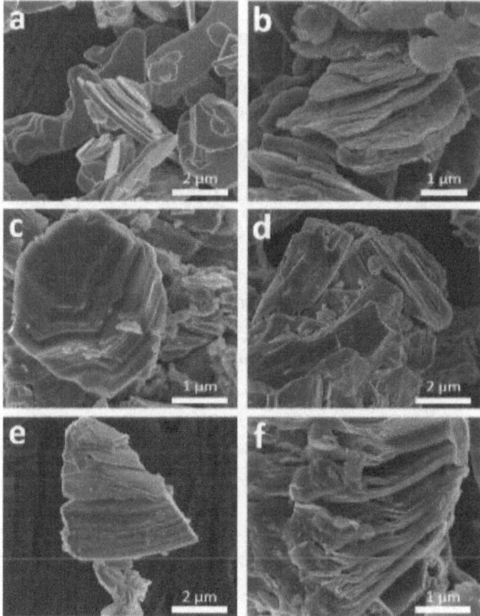

Figure 5.2. SEM micrographs of the parent layered carbides and multilayer HF etched MXenes. (a) Mo_2Ga_2C, (b) multilayer Mo_2CT_x, (c) Mo_2TiAlC_2, (d) multilayer $Mo_2TiC_2T_x$, (e) $Mo_2Ti_2AlC_3$, (f) multilayer $Mo_2Ti_2C_3T_x$. Reprinted with permission.[19] Copyright 2017 American Chemical Society.

Figure 5.3(a) shows a digital photograph of Mo_2CT_x, $Mo_2TiC_2T_x$, and $Mo_2Ti_2C_3T_x$ freestanding papers, which were peeled off from a porous membrane after vacuum-assisted filtration and cut into 1×1 cm^2 squares for thermoelectric property measurements. Those freestanding paper samples are usually flexible in the pristine state without the use of binder additives, as shown in **Figure 5.3(b)**. The flexibility can be altered upon annealing unless the nanosheets are supported by a flexible substrate. To achieve both high electrical conductivity and flexibility of MXene paper in pristine or mild dried state, cation intercalation can be used instead of organic intercalation as demonstrated in $Ti_3C_2T_x$ MXene.[14, 22] **Figure 5.3(c-h)** show cross-sectional SEM micrographs of the as-synthesized Mo-based MXene papers (pristine) and after the thermoelectric measurements from room temperature (RT) to 800 K (annealed). The typical layered structure due to the restacking of MXene flakes is clearly observed in the pristine MXene papers. The cross-sectional SEM images of the annealed MXene papers show more compact structures, which lead to increased electrical conductivity, as will be discussed later.

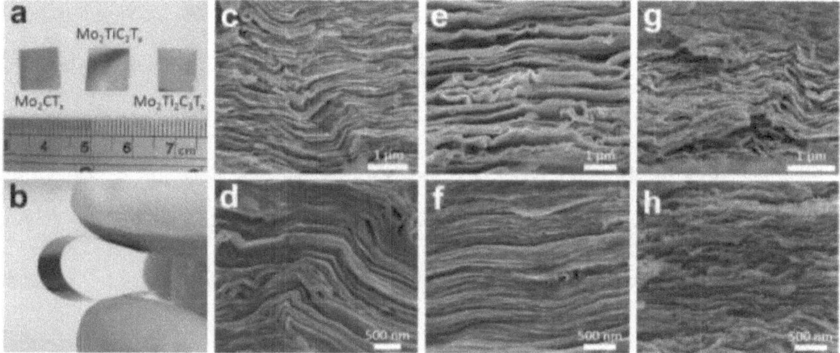

Figure 5.3. (a) Digital photographs of Mo-based MXene papers. (b) Demonstration of the flexibility of freestanding $Mo_2Ti_2C_3T_x$ paper in pristine state. (c-h) Cross-section SEM micrographs of Mo-based MXene papers; (c) pristine Mo_2CT_x, (d) annealed Mo_2CT_x, (e) pristine $Mo_2TiC_2T_x$, (f) annealed $Mo_2TiC_2T_x$, (g) pristine $Mo_2Ti_2C_3T_x$, (h) annealed $Mo_2Ti_2C_3T_x$. Reprinted with permission.[19] Copyright 2017 American Chemical Society.

X-ray diffraction (XRD) patterns of the parent layered carbides with corresponding MXene papers in pristine and annealed states are presented in **Figure 5.4**. Pristine MXene papers (middle patterns) show a large shift of the (002) peak toward lower angles compared to corresponding parent carbide precursor phases (bottom patterns). This result suggests an increase in the c-lattice parameter (c-LP) in the MXene papers due to the surface functional groups replacing the 'A-element' layers, as well as the presence of intercalated TBA^+ and water molecules between the MXene nanosheets.[167] After annealing the MXene papers at 800 K (top patterns), the (002) peak shifts back to higher angles, which can be explained by a decrease in the c-LP due to the removal of the intercalated molecules, OH and some other terminations.[40] For Mo_2Ga_2C and its MXene (pristine and annealed, **Figure 5.4(a)**), the (002) peaks are centered at 9.76°, 5.24°, and 11.17°, respectively. The corresponding c-LPs are 18.1 Å, 33.7 Å, and 15.8 Å. The interlayer spacing between Mo_2C layers in pristine Mo_2CT_x compared to Mo_2Ga_2C increased by ~ 7.8 Å. In Mo_2Ga_2C, two atomic layers of Ga separate the Mo_2C layers. After etching and delamination, these Ga layers are replaced by surface terminations and intercalated TBA^+ and water molecules. After annealing the Mo_2CT_x paper, the interlayer spacing decreased by ~ 8.9 Å, due to the removal of intercalated molecules and cations. Mo_2CT_x with no intercalated molecules has been calculated to have similar c-LP.[168] Considering the synthesis process described above, one observes a large change in the interlayer distance, which is due to the intercalation (pristine MXene) and removal of organic molecules (annealed MXene).

A similar trend can be found in Mo_2TiAlC_2 (**Figure 5.4(b)**), where the c-LP value first increases from 18.5 Å for the precursor (Mo_2TiAlC_2) to 37.8 Å after etching the Al layers and intercalation, and decreases to 24.5 Å due to the removal of intercalated organic and water molecules during heating to 800 K. In the case of $Mo_2Ti_2AlC_3$ (**Figure 5.4(c)**), the initial c-LP value is 23.4 Å based on the (004) peak position at 15.13° for the $Mo_2Ti_2AlC_3$, and it increases to 42.6 Å for the pristine $Mo_2Ti_2C_3T_x$ paper and decreases to 27.6 Å after annealing. The crystallographic ionic diameter of TBA^+ is known to be around 9.9 Å, however a minor distortion/conformation of butyl chains allows the intercalation into 8 Å interlayer space.[169] Thus, the observed shifts for the basal plane related peak positions correlate with the intercalation and de-intercalation of TBA^+ and water molecules.

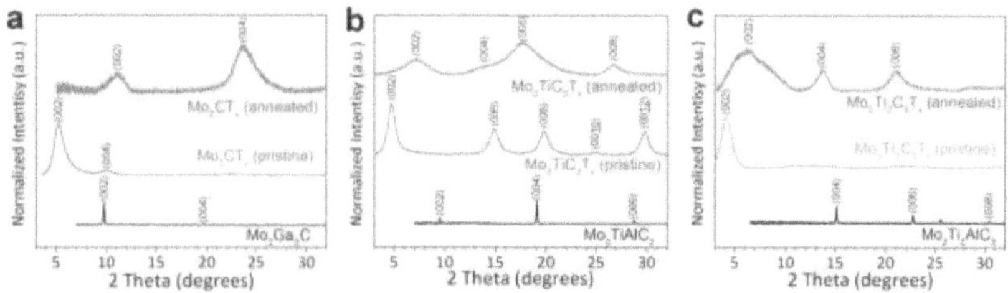

Figure 5.4. X-ray diffraction patterns of the parent layered carbide phases and corresponding pristine and annealed MXene papers. (a) Mo_2Ga_2C and Mo_2CT_x (b) Mo_2TiAlC_2 and $Mo_2TiC_2T_x$, (c) $Mo_2Ti_2AlC_3$ and $Mo_2Ti_2C_3T_x$. There are no XRD peaks below 8 degree in patterns of precursors. Reprinted with permission.[19] Copyright 2017 American Chemical Society.

Figure 5.5 shows the Raman spectra of three Mo-based layered carbide precursors (bottom) and their corresponding pristine (middle) and annealed MXenes (top). There are clear changes in the Raman spectra after etching that show disappearance and/or shift of the Raman active vibration modes related to M-A vibrations, and appearance of new modes which may be related to surface functionalization and/or additional modes becoming active in 2D layers, when the constraint of M-A bonding in the 3D structure has been removed. Weak carbon bands appeared after annealing, suggesting either formation of disordered carbon due to unavoidable carbon contamination such as carbon dioxide, or some disproportionation of MXene possibly due to partial loss of surface terminations.[170] All peaks of MXene stay in place and peak shifts are minor, indicating that the structure of MXene was retained after annealing and the reported properties are representative of the MXenes, not their decomposition products. Our experimental Raman spectra are in good agreement with multiple simulation studies using O termination.

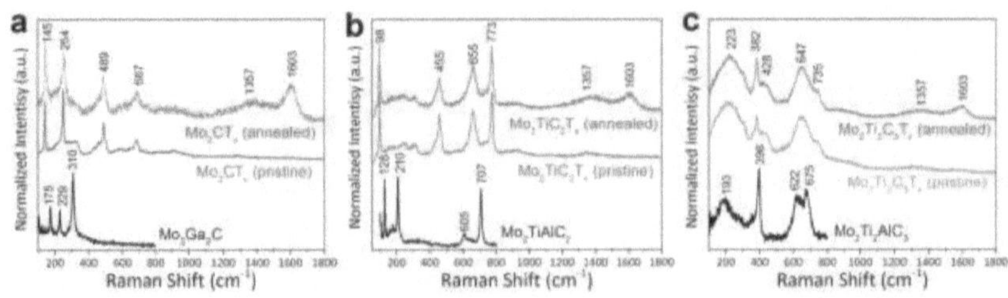

Figure 5.5. Raman spectra of the parent layered carbide phases and corresponding pristine and annealed MXene papers. (a) Mo_2Ga_2C and Mo_2CT_x (b) Mo_2TiAlC_2 and $Mo_2TiC_2T_x$, (c) $Mo_2Ti_2AlC_3$ and $Mo_2Ti_2C_3T_x$. There are no Raman peaks above 800 cm^{-1} in the spectra of precursors. Reprinted with permission.[19] Copyright 2017 American Chemical Society.

For the precursor materials (bottom spectra), it has been reported that the low frequency modes can be attributed to M- and A- element layers related vibrational modes, while the higher frequency modes are dominated by 'M-X' bonds.[84, 171] It should be noted that the latter case can only observed in $M_{n+1}AX_n$ phases with n > 2.[84] Thus, one may expect elimination of the 'A-element' layers related peaks from the MXene spectra. The Raman spectra of pristine Mo-based MXene papers (middle spectra) differ from that of the parent phases, suggesting successful exfoliation and separation of 2D MXene flakes. At low wavenumbers (~350 cm^{-1}), the vibration modes dominated by metal atoms are shifted or diminished. Interestingly, additional Raman peaks have been observed, which are not common to their parent phases. This is due to the additional vibrational modes dominated by the surface functional groups. The vibration modes related to carbon atoms, at high wavenumber (600 – 800 cm^{-1}), are also affected by surface functional groups replacing the A-element layers. It is reported that different 'A-element' layers can affect this 'M-X' bond related vibration modes.[171] The annealed MXenes (top spectra) show more or less similar Raman spectra to the pristine MXene, along with additional two peaks at around 1357 and

1603 cm^{-1} which are attributed to D- and G-bands of graphitic carbon, respectively.[170] In the cases of oxidized Ti_3C_2 (d_{c-c}=3.0Å, 1596cm^{-1})[172] and oxidized Nb_2C (d_{c-c}=3.1Å, 1614cm^{-1}),[170] the observed G-bands are slightly shifted to higher frequency compared to that of graphite (d_{c-c}=1.4Å, 1580cm^{-1}). These are the result of disproportionation reaction of MXenes where it forms oxide and amorphous carbons. It should be noted that the 2D-band was not observed (**Figure 5.6**) because no continuous graphene sheets were formed. The similar absence of 2D- band is reported in the case of graphene oxide.[173]

For Mo_2CT_x MXene (**Figure 5.5(a)**), the observed Raman peak centers match the theoretical reports on phonon dispersion and Raman-active frequencies of various M_2CT_x.[8] The low frequency peaks centered at 145 and 254 cm^{-1} are attributed to the E_{2g} and A_{1g} modes, respectively, that are dominated by Mo atoms. These two sharp peaks are shifted from the 175 and 229 cm^{-1} observed for the parent Mo_2Ga_2C phase. The other peak at 310 cm^{-1} in the Mo_2Ga_2C spectra is diminished and slightly shifted to a higher frequency in Mo_2CT_x, which is related to the A_{1g} mode involving displacement of Mo atoms perpendicular to the basal plane where the nearby Ga atoms vibrate in opposite direction of Mo atoms. (Detailed interpretation for Mo_2Ga_2C phase is available from Chaix-Pluchery *et al.*[174]) The corresponding vibration mode in Mo_2CT_x is likely affected by the surface functional groups, which form a double layer between MXene flakes replacing the Ga layers in Mo_2Ga_2C. The other two peaks of Mo_2CT_x centered at around 489 and 687 cm^{-1} are additional E_{2g} and A_{1g} modes, respectively, caused by the presence of surface functional groups. All four peaks agree well with the reported phonon dispersion of Mo_2CO_2, where oxygen atoms are located in the empty sites of three topmost Mo atoms, and directly above the carbon atoms in the below layer.[8, 52]

In the case of $Mo_2TiC_2T_x$ MXene (**Figure 5.5(b)**), our experimental Raman spectra are in good agreement with the simulation studies of $Mo_2TiC_2O_2$,[50, 51, 175] which uses the same surface functional group model described above, along with spin-orbit coupling. A sharp peak at 98 cm^{-1} can be attributed to the E_g mode dominated by Mo atoms, and is shifted from 128 cm^{-1} of the parent Mo_2TiAlC_2 MAX phase. The other peak of Mo_2TiAlC_2 at 210 cm^{-1}, which can be assigned to the A_{1g} mode involving the relative displacement of Mo and C atoms along the c-axis, is significantly diminished in $Mo_2TiC_2T_x$ MXenes. An additional peak centered at 455 cm^{-1} is observed in $Mo_2TiC_2T_x$ MXenes, which can be attributed to the oxygen dominated vibrational mode along the basal plane. The other two higher frequency mode at 655 and 773 cm^{-1} are dominated by carbon atoms parallel (E_g) and perpendicular (A_g) to the basal plane, respectively.

For $Mo_2Ti_2C_3T_x$ MXene (**Figure 5.5(c)**), relatively broad spectra were observed, which is similar to the parent $Mo_2Ti_2AlC_3$ MAX phase. It should be noted that the M_4AX_3 structures are known to have 10 Raman-active modes; however we observed three broad bands, with one relatively sharp peak. Due to the instrumental resolution, it is possible that the Raman modes are so close to each other that they appear as one broad band.[171] Defects, bending of nanosheets and diverse surface terminations certainly contribute to peak broadening. The sharp peak of $Mo_2Ti_2AlC_3$ at 398 cm^{-1} is the A_{1g} mode dominated by the displacement of metal atoms along the c-axis. This mode seems to be shifted to 382cm^{-1} in $Mo_2Ti_2C_3$

MXene with decreased peak intensity. It should be noted that the similar vibrational modes of Mo_2Ga_2C (310 cm^{-1}) and Mo_2TiAlC_2 (210 cm^{-1}) became significantly diminished in Mo_2CT_x and $Mo_2TiC_2T_x$, which is likely due to the damped volume changes by the surface functional groups. For $Mo_2Ti_2C_3T_x$, which has four layers of metal atoms, a smaller volumetric restriction is expected due to the surface functional groups as compared to Mo_2CT_x and $Mo_2TiC_2T_x$, thus this vibration mode can only be observed in $Mo_2Ti_2C_3T_x$. The observed broad band at a low frequency is dominated by metal atoms, and an additional shoulder peak at 428 cm^{-1} is probably due to the oxygen terminated surface. The other two higher frequency modes at 647 and 735 cm^{-1} are due to the displacement of carbon atoms, parallel and perpendicular to the basal plane, respectively. These two carbon related modes are more broad and shifted to lower frequencies compared to $Mo_2TiC_2T_x$, which is possibly due to multiple carbon states in $Mo_2Ti_2C_3T_x$; this includes carbon atoms located in either Mo_3Ti_3C (first and third carbon layer) or Ti_6C (second carbon layer) octahedrons. This indicates a relatively weak Ti-C bond compared to Mo-C, resulting in red-shifted vibrational modes. The observed Raman spectra of $Mo_2Ti_2C_3T_x$ also match the reported phonon dispersion curves of $Mo_2Ti_2C_3O_2$ using the same model for the surface termination described above and including quantum spin Hall effect.[176]

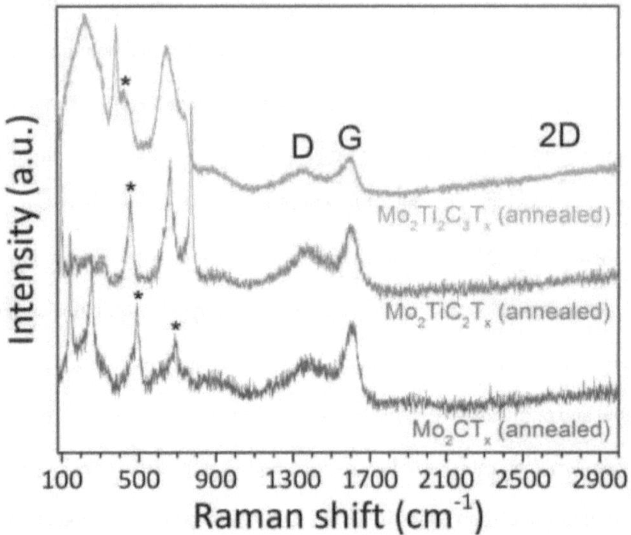

Figure 5.6. Raman spectra of annealed Mo-based MXene papers, showing the absence of 2D-band near 2700 cm^{-1}, despite the presence of D and G bands of graphitic carbon. Asterisks denote the peaks related to surface terminated oxygen. Reprinted with permission.[19] Copyright 2017 American Chemical Society.

5.3 Thermoelectric Properties

Temperature dependent thermoelectric properties of the Mo-based MXenes are presented in **Figure 5.7(a-c)**. All three MXene papers are highly resistive in the pristine state near RT, but their electrical conductivity increases rapidly during the first heating cycle, with three identifiable regimes: (1) Removal of water molecules gradually enhances the conductivity at low temperatures (300-500 K). (2) Thermal decomposition of TBA$^+$

intercalant at around 500-600 K. The onset of rapid conductivity increase is well matched to the reported thermal decomposition temperature of TBA$^+$ intercalant in graphite, which is around 480 K.[169] (3) Above 600 K, the conductivity increases slowly which can be ascribed to the decomposition of remaining TBA$^+$ and the possible thermal elimination of surface termination groups. It should be noted that the observed temperature dependent electrical conductivity during the 1st heating cycle is the sum of both irreversible changes due to thermal de-intercalation, and reversible thermal response of the material itself, where the former factor is more dominant. The MXene flakes are likely to undergo re-arrangement throughout the thermal heating process described above, where the interlayer spacing decreases.

In order to confirm the above hypothesis about the cause of the conductivity change during the first heating cycle in MXenes, thermal gravimetric analysis (TGA) was carried out along with differential scanning calorimetry (DSC) for $Mo_2TiC_2T_x$ paper, as shown in **Figure 5.7(d)**. An overall weight loss of around 8% was observed from RT to 800 K under Ar ambient. The quantity of intercalated water molecules in pristine $Mo_2TiC_2T_x$ is estimated around 1.6 wt% based on the weight loss below 500 K.[177] Most of weight loss (5.3 wt%) between 500-650 K can be attributed to the thermal de-intercalation of TBA$^+$, and this temperature range matches the one in which electrical conductivity of MXenes shows a significant increase. This result suggests that the MXene flakes in the pristine paper are separated by large intercalant molecules, but come in closer contact that enhances the carrier transport due to the thermal de-intercalation process at around 500-600 K. These changes, also supported by XRD studies (**Figure 5.4**), are irreversible and are not observed in the 2nd thermal cycle of TGA-DSC, nor during the 2nd thermal cycle of thermoelectric properties measurement (**Figures 5.8 - 5.10**). At above 650 K, only ~1 wt% of weight loss was observed, while the electrical conductivity slowly increased. Despite the small changes in both weight and electrical conductivity above 650 K, relatively large endothermic heat flow was observed during both the first and second thermal cycle. Considering that the weight change in second thermal cycle is nearly zero, the heat could be absorbed due to the removal of surface functional groups, as earlier observation for $Ti_3C_2T_x$.[40] The c-LP of $Ti_3C_2T_x$ without intercalated water molecule decreased by only 0.5 Å after annealing up to 773 K under vacuum.[178] Since the surface functional groups are known to accept electrons from MXenes,[175] removal of surface terminations can be another contributor to the increase in electrical conductivity.

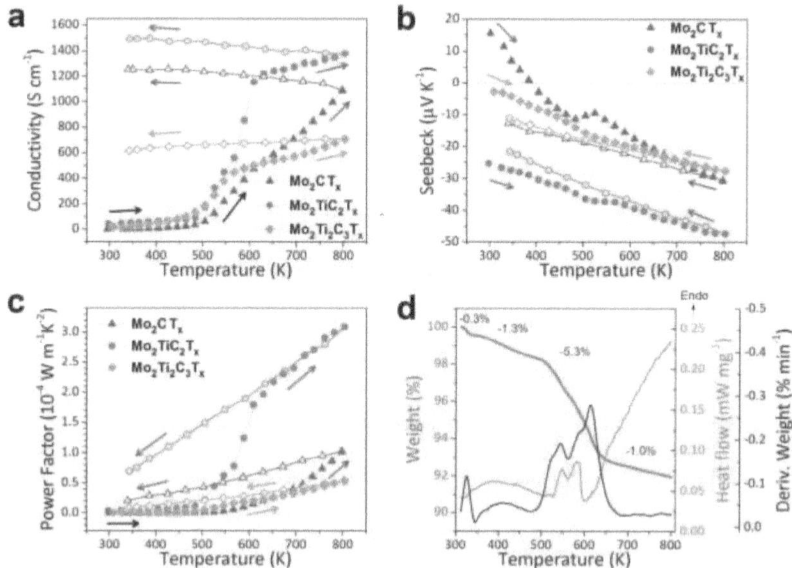

Figure 5.7. (a-c) Temperature dependent thermoelectric properties of Mo-based MXene papers during the first thermal cycle. (a) Electrical conductivity, (b) Seebeck coefficient, and (c) thermoelectric power factor. The heating cycle and subsequent cooling cycle are marked by filled symbols and open symbols, respectively. The blue triangle, green circle, orange diamond symbols represent Mo_2CT_x, $Mo_2TiC_2T_x$, and $Mo_2Ti_2C_3T_x$, respectively. (d) TGA curve (green) with first derivative of weight (black) and DSC curve (red) of $Mo_2TiC_2T_x$ for the first heating cycle. Reprinted with permission.[19] Copyright 2017 American Chemical Society.

The electrical conductivity of pristine Mo_2CT_x measured before annealing is 0.8 S cm^{-1}, but increases to 1252 S cm^{-1} at the end of the first thermal cycle (heating and then cooling to near RT) during thermoelectric measurements. This result is in agreement with literature, where the RT electrical conductivity of Mo_2CT_x was shown to increase by two orders of magnitude after drying at 120°C for 18 hours under vacuum.[44] The conductivity values of pristine (annealed) $Mo_2TiC_2T_x$ and $Mo_2Ti_2C_3T_x$, are 44 S cm^{-1} (1494 S cm^{-1}) and 15 S cm^{-}

^{1} (614 S cm^{-1}), respectively. **Table 5.1** shows a summary of the Hall-effect measurement results. Pristine Mo-based MXene papers typically have carrier density of $\sim 10^{20}$ cm^{-3} with a Hall mobility of $10^{-2} - 10^{-1}$ cm^2 V^{-1} s^{-1}. Annealed Mo-based MXene papers exhibit one order of magnitude higher carrier concentration with Hall mobility of few cm^2 V^{-1} s^{-1}. The changes in conductivity during the first annealing are attributed to the decreased interlayer spacing due to thermal de-intercalation and partial de-functionalization of MXenes. Once the conductivity stabilizes after the first heating cycle, Mo$_2$CT$_x$ and Mo$_2$TiC$_2$T$_x$ exhibit metallic behavior while Mo$_2$Ti$_2$C$_3$T$_x$ shows semiconductor-like behavior, based on the slope of temperature dependence of conductivity in the stabilized (annealed) state. These characteristics are related to the carrier concentration and may have also been affected by the stacking structure, where annealed Mo$_2$CT$_x$ and Mo$_2$TiC$_2$T$_x$ show well aligned nanosheets (**Figure 5.3(d,f)**), while annealed Mo$_2$Ti$_2$C$_3$T$_x$ films look less compacted (**Figure 5.3(h)**). Larger thickness of Mo$_2$Ti$_2$C$_3$T$_x$ nanosheets could make their rearrangement and film densification more difficult, but we did not find a good way to quantify those differences.

Table 5.1. Hall-effect measurement results of Mo-based MXene papers at RT.

Sample	n (1/cm^3) $^{a)}$	μ (cm^2/V·s) $^{b)}$	Type
Mo$_2$CT$_x$ (pristine)	2.84×10^{20}	0.016	p
Mo$_2$TiC$_2$T$_x$ (pristine)	8.31×10^{20}	0.323	n
Mo$_2$Ti$_2$C$_3$T$_x$ (pristine)	2.01×10^{20}	0.451	n
Mo$_2$CT$_x$ (annealed)	4.10×10^{21}	1.86	n
Mo$_2$TiC$_2$T$_x$ (annealed)	3.04×10^{21}	2.85	n
Mo$_2$Ti$_2$C$_3$T$_x$ (annealed)	1.83×10^{21}	2.05	n

a) Carrier concentration (n), b) Hall mobility (μ).

The Seebeck coefficient of these MXenes were simultaneously measured as shown in **Figure 5.7(b)**. All three MXenes show n-type behavior, as indicated by the negative values of the Seebeck coefficients. Unlike the rapid change in electrical conductivity, the Seebeck coefficient does not show a drastic transition; instead it undergoes a slow change during the first heating cycle. This observation is another evidence that supports the fact that the rapid increase in the electrical conductivity above ~500 K (**Figure 5.7(a)**) is likely due to the enhanced electrical network between MXene nanosheets rather than irreversible changes of each MXene flakes. As the nanosheets get better connection during the first heating cycle, the energy barrier between MXene flakes will spontaneously decrease, hence a larger numbers of domains will contribute to its overall Seebeck coefficient. Particularly, Mo_2CT_x samples shows an interesting transition from p-type to n-type upon thermal annealing, as confirmed by both Hall-effect measurement and thermoelelctric test. The spontaneous transition of its Seebeck coefficient seems to be a result of a decreased energy barrier at the edges of MXene flakes during the first heating cycle.

Just like electrical conductivity, the Seebeck coefficient of all three MXenes reaches a steady-state value and remains stable after the first heating cycle, as shown in **Figures 5.8 – 5.10**. The temperature dependent Seebeck coefficient is quite linear with temperature, indicating absence of any magnetic transition. $Mo_2Ti_2C_3T_x$ exhibits the smallest Seebeck coefficient, among the three samples, of -27.5 $\mu V \ K^{-1}$ at 803 K, while Mo_2CT_x exhibits -30.5 $\mu V \ K^{-1}$ at 798 K. These two compositions have similar temperature dependent Seebeck coefficient. Mo_2CF_2 has been theoretically predicted to be semiconducting with a large Seebeck coefficient (> 100$\mu V \ K^{-1}$ at 400 K) and a negative correlation between

Seebeck coefficient and temperature.[57] However, our experimental results show the opposite. This suggests that the prepared Mo_2CT_x may not be dominated by fluorine termination, despite the fact that we used an HF etching method. It has been reported that delamination of Mo-containing MXenes with TBAOH reduces the fluorine content.[38] In contrast, $Mo_2TiC_2T_x$ exhibited a larger Seebeck coefficient of -47.3 μV K^{-1} at 803 K, along with the largest slope. It is worth to note that the observed Seebeck coefficient of $Mo_2TiC_2T_x$ MXene paper is higher than the theoretically calculated value of its parent Mo_2TiAlC_2 MAX phase.[179]

Due to its high conductivity (1380 S cm^{-1} at 803 K) and large Seebeck coefficient (-47.3 μV K^{-1} at 803K), $Mo_2TiC_2T_x$ shows the largest thermoelectric power factor of 3.09×10^{-4} W m^{-1} K^{-2} at 803 K among the MXenes studied here, as shown in **Figure 5.7(c)**. $Mo_2TiC_2T_x$ also shows nearly one order of magnitude higher power factor near RT compared to chemically exfoliated WS_2, which has been demonstrated as a wearable thermoelectric generator.[180] This superior thermoelectric power factor of $Mo_2TiC_2T_x$ can be attributed to its electronic band structure. One possible explanation is the so-called "pudding-mold" type band structure found in high performance thermoelectric materials such as Na_xCoO_2 and $CuAlO_2$.[181, 182] The band near the Fermi level can be flatter depending on the ratio of hopping integrals to the first, second, and third nearest neighbor sites on the triangular lattice,[182, 183] where the peculiar shape band structure can result multiple Fermi surfaces that can contribute to the conductivity overcoming the decrease in Seebeck coefficient. Such variable range hopping behavior can be related to the geometrical frustration in triangular lattice with the antiferromagnetic spin interaction. However, more theoretical work is needed to confirm this hypothesis. According to the simulation studies on the electronic band structure of Mo-based MXenes with oxygen termination,[50, 52, 176] $Mo_2TiC_2O_2$ certainly shows a much larger portion of flat band near the Fermi level than Mo_2CO_2 and $Mo_2Ti_2C_3O_2$, which can be the origin of coexistence of metallic conductivity and a relatively larger Seebeck coefficient. Due to the highly conductive nature of MXenes, hybrid materials with outstanding thermoelectric performance may be produced by combining MXenes with large band gap semiconductors to increase the overall Seebeck coefficient and/or polymers to decrease the thermal conductivity.[184, 185]

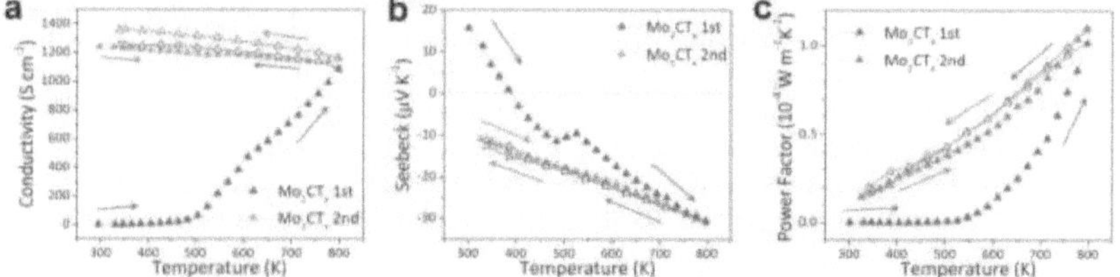

Figure 5.8. Temperature dependent thermoelectric properties of Mo_2CT_x paper. The first and second thermal cycles are presented by blue and red symbols, respectively. The heating cycle and subsequent cooling cycle are marked by filled symbols and open symbols, respectively. Reprinted with permission.[19] Copyright 2017 American Chemical Society.

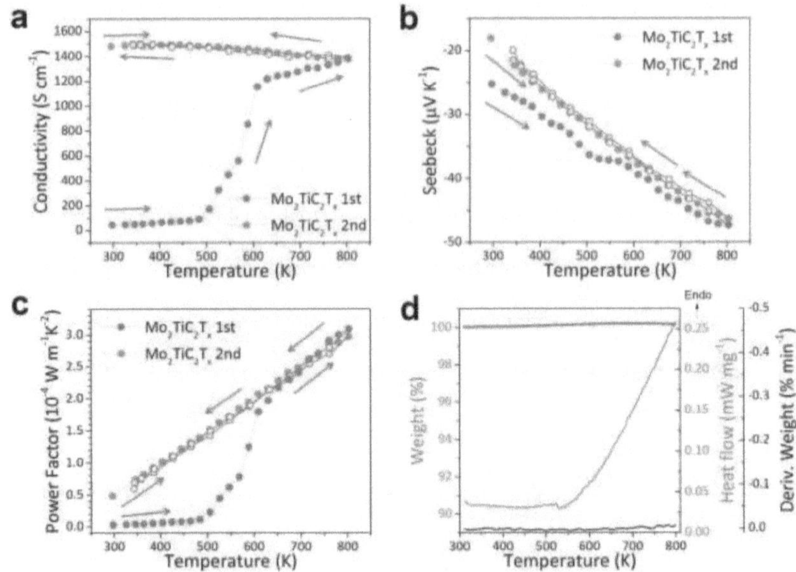

Figure 5.9. (a-c) Temperature dependent thermoelectric properties of $Mo_2TiC_2T_x$ paper. The first and second thermal cycles are presented by green and red symbols, respectively. (a) Electrical conductivity, (b) Seebeck coefficient, and (c) thermoelectric power factor. The heating cycle and subsequent cooling cycle are marked by filled symbols and open symbols, respectively. (d) TGA curve (green) with first derivative of weight (black) and DSC curve (red) of $Mo_2TiC_2T_x$ for the second heating cycle. No weight loss is observed in the second heating cycle. Reprinted with permission.[19] Copyright 2017 American Chemical Society.

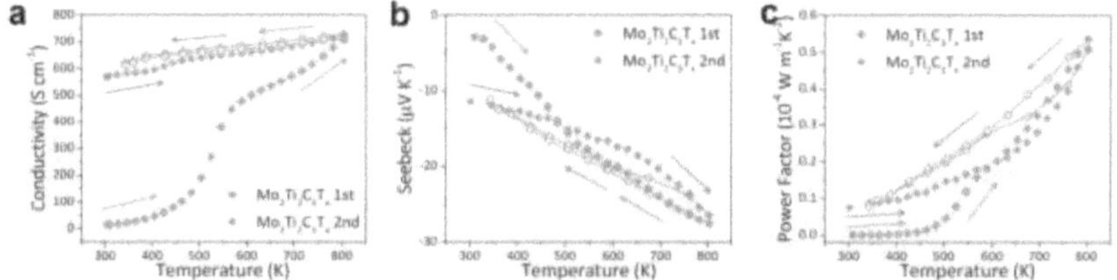

Figure 5.10. Temperature dependent thermoelectric properties of $Mo_2Ti_2C_3T_x$ paper. The first and second thermal cycles are presented by orange and grey symbols, respectively. The heating cycle and subsequent cooling cycle are marked by filled symbols and open symbols, respectively. Reprinted with permission.[19] Copyright 2017 American Chemical Society.

5.4 Conclusions

In summary, we have reported the first experimental measurements of the temperature dependent thermoelectric properties of Mo-based MXenes up to 800 K. A combination of a high electrical conductivity (1380 S cm^{-1} at 803 K) and a relatively large Seebeck coefficient (-47.3 µV K^{-1} at 803K) has been found for $Mo_2TiC_2T_x$, with its thermoelectric power factor reaching 3.09×10^{-4} W m^{-1} K^{-2} at 803 K. The conductivity of all MXenes studied in this work shows a rapid increase above ~500 K, which is due to de-intercalation of water and organic molecules, as well as partial loss of functional groups resulting in a decrease of interlayer spacing and improved contacts between the MXene nanosheets. Raman spectroscopy analysis revealed that Mo-based MXenes are thermally robust upon annealing at up to 800 K under Ar/H$_2$ ambient. Further enhancement in thermoelectric performance of MXenes can be achieved by additional doping, controlling the surface functional groups, and/or hybridization with semiconductor materials or polymers.

5.5 Experimental Section

Synthesis of MXene freestanding papers: To synthesize Mo_2CT_x,[44] $Mo_2TiC_2T_x$,[38] and $Mo_2Ti_2C_3T_x$[38] MXenes, 2g of Mo_2Ga_2C powder was slowly added into 80 ml of 25% aqueous hydrofluoric acid (HF), and for Mo_2TiAlC_2 and $Mo_2Ti_2AlC_3$ MAX phases 20 ml of 50% HF were used. The precursor carbides were slowly added into the acidic solutions which were cooled by ice-bath to minimize the localized heating. The mixtures were then stirred with a magnetic Teflon stir bar at 55°C for 160 h, 48 h, and 96 h, respectively. After etching, the mixtures were washed several times until the pH reached around 6. For each washing step, 40 ml of deionized (DI) water was added into centrifuge tubes, hand shaken for 1 min, centrifuged at 5000 rpm for 2 min, and then discarded the supernatant. The solid sediments were kept in a vacuum oven at RT overnight for drying. 1g of dried multilayer MXene powder was further re-dispersed into 10 ml of 54-56% tetrabutylammonium hydroxide (TBAOH) solution for intercalation, and stirred for 4 h (Mo_2CT_x, $Mo_2TiC_2T_x$) or 18 h ($Mo_2Ti_2C_3T_x$) at RT. The resulting mixtures were washed two times, and sediments were re-dispersed into DI water followed by ultra-sonication for an hour. The mixtures were later centrifuged at 5000 rpm for 1 hour, and the resulting supernatants were carefully collected. The suspension solutions were vacuum filtrated through a porous membrane (3501, surfactant-coated PP, Celgard, USA), and the vacuum filtrated papers were carefully peeled off for the experiment. 50% HF and 54-56% TBAOH were purchased from Aldrich.

Thermoelectric properties measurement: Electrical conductivity and Seebeck coefficient of MXene paper were measured in the in-plane direction under dynamic flow of Ar/H_2 reducing ambient. The measurements were done in the temperature range of 300-

800 K, using a commercially available thermoelectric tester RZ2001i (Ozawa Science Co. Ltd., Japan). At each temperature, electrical conductivity were measured under thermal equilibrium by 4-point linear probe method. After that, a small thermal gradient was generated by local cooling of one side of sample (local cooling was done by air circulation through one of the quartz sample holder pipe). Among the four linear probes, two of the inner probes are thermocouple-probes, so the temperature difference (ΔT) and the voltage difference (ΔV) can be measured across the sample. Five different data points were collected at each temperature, and the Seebeck coefficient could be calculated by finding the linear fit slope ($\Delta V/\Delta T$). As the measured Seebeck coefficient is the difference between the sample and the platinum electrode ($S_{measured} = S_{sample} - S_{electrode}$), corresponding Seebeck coefficient of Pt at the temperature was accounted for to find the Seebeck coefficient of the sample. Thermal history for the measurement is shown in **Figure 5.11**. Further details on the measurement can be found elsewhere.[186]

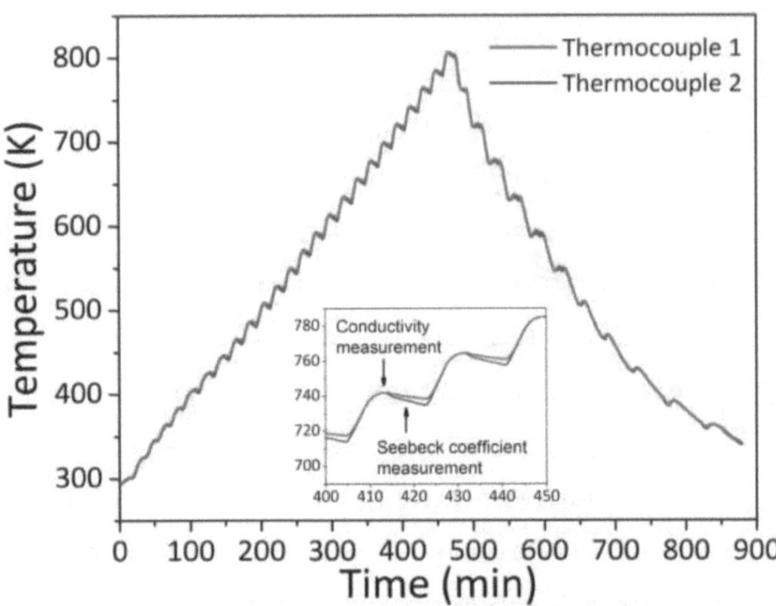

Figure 5.11. Thermal history of temperature dependent thermoelectric measurement. The electrical conductivity was first measured under thermal equilibrium, followed by the Seebeck coefficient measurement while generating a small localized thermal gradient across the sample. Reprinted with permission.[19] Copyright 2017 American Chemical Society.

Characterization: X-ray diffraction (XRD) patterns were collected by a Bruker diffractometer (D8 advance, AXS system, Germany) with Cu Kα radiation, $\lambda = 1.5406$ Å. For the precursor carbides, small amounts of powder were loaded into a small cavity of the sample holder and pressed using a glass slide. For MXene, a single slice of MXene paper was placed on a zero-diffraction plate. Cross-section morphologies were observed by FE-SEM (Nova Nano 630, FEI, USA). The Raman spectra were collected by a micro-Raman spectrometer (LabRAM Aramis, Horiba, Japan) equipped with a long working distance

×50 objective lens (LMPLFLN50X, Olympus, Japan) using 473 nm excitation (04-01 series, Cobalt blues™, Sweden) at RT. Thermal analysis (TGA-DSC) was carried out by STA 449 F1 Jupiter (Netzsch, Germany) with the heating rate of 5 K min^{-1} under Ar ambient, using Al_2O_3 crucibles. Carrier concentration and Hall mobility were measured by a Hall-effect measurement system (Lake Shore 7700A) using van der Pauw technique at RT under a magnetic field range of ± 5, 10, 15, and 20 kG. Ohmic contact was achieved by using fast drying silver paste.

Chapter 6

Summary and Perspectives

6.1 Summary

The main conclusion of this book work is that MXenes can be used as alternative conductive electrodes in electronic device applications. The unique property combination of hydrophilic surface and metallic conductivity of $Ti_3C_2T_x$ MXene promises a great potential. They could be formed into flexible, transparent, and conductive electrode by simple solution-process, and their tunable work function is attractive as contact material for semiconductors. By exploring the solution process of MXene thin films and their photolithographic patterning process, transistors with oxide semiconductors and quantum dot semiconductors have been investigated with lift-off and dry-etch methods, respectively.

The first utilization of MXene as electronic contact material has been investigated for oxide thin film transistors. MXene electrodes are patterned by lift-off process, where MXene suspension was spray coated over the patterned photoresist layer. Metallic $Ti_3C_2T_x$ MXene with work function of 4.60 eV can make good electrical contact with both n-type ZnO and p-type SnO semiconductors, with negligible band offsets. Consequently, we have fabricated both n-type ZnO and p-type SnO thin film transistors (TFTs) entirely using large-area MXene electrical contacts, including gate, source, and drain. The n- and p-type TFTs show balanced performance, including field-effect mobilities of 2.61 and 2.01 cm^2 V^{-1} s^{-1}, switching ratios of 3.6×10^6 and 1.1×10^3, respectively. Based on the balanced TFT performance, complementary metal oxide semiconductor (CMOS) inverters are further demonstrated. The CMOS inverters show large voltage gain of 80, excellent noise margin of 3.54 V, which is 70.8% of the ideal value. Moreover, the operation of CMOS inverters is shown to be very stable under 100 Hz square waveform input.

A fully solution-processed, large-area, electrical double layer transistors (EDLTs) is presented by employing lead sulfide (PbS) colloidal quantum dots (CQDs) as active channels and $Ti_3C_2T_x$ MXene as electrical contacts (including gate, source, and drain). The MXene contacts are patterned by standard photolithography and plasma-etch techniques for the first time and successfully integrated with CQD films. This is to overcome the limitation of lift-off method which is remaining sidewall-like residue due to the conformal spray-coating process. The large surface area of CQD film channels is effectively gated by ionic gel, resulting in a EDLTs exhibiting remarkable performances. A large electron saturation mobility of 3.32 cm^2 V^{-1} s^{-1} and current modulation of 1.87×10^4 operating at low driving gate voltage range of 1.25 V, with negligible hysteresis are achieved. The relatively low work function of $Ti_3C_2T_x$ MXene (4.4 eV in ultrahigh vacuum) compared to vacuum-evaporated noble metals such as Au and Pt makes them a suitable contact material for n-type transport in iodide-capped PbS CQD films with a LUMO level of ~ 4.14 eV. Moreover, this work demonstrates the first utilization of negative surface charges of MXene for the accumulation of cations at the lower gate bias, achieving a threshold voltage as low as 0.36 V.

Three compositions of Mo-based MXenes (Mo_2CT_x, $Mo_2TiC_2T_x$, and $Mo_2Ti_2C_3T_x$) have been synthesized and processed into free-standing binder-free papers by vacuum-assisted filtration, and their electrical and thermoelectric properties are measured. Upon heating to 800 K, these MXene papers exhibit high conductivity and n-type Seebeck coefficient. The thermoelectric power reaches 3.09×10^{-4} W m^{-1} K^{-2} at 803 K for the $Mo_2TiC_2T_x$ MXene. While the thermoelectric properties of MXenes do not reach that of the best materials, they exceed their parent ternary and quaternary layered carbides. $Mo_2TiC_2T_x$ shows the highest electrical conductivity in combination with the largest Seebeck coefficient of the three 2D materials studied that had been theoretically predicted to have excellent thermoelectric performance among many other MXenes.

6.2 Challenges and Perspectives on Future Work

Detailed and controllable studies are desired on the impact of such modification on MXenes' fundamental properties and MXene-based devices. Plasma, chemical, and thermal treatment present possible approaches to modify MXene surfaces. These treatments can be used to tailor the MXene work function, which then enables its utilization to make devices such as Schottky junction or Ohmic contacts for transistors, diodes, solar cells, and photodetectors. The surface terminations may affect other properties such as plasmonic behavior, thermoelectric properties, and photothermal properties of MXenes.

For nearly 10 years of investigation of MXenes, it is still limited to be obtained from solution-based chemistry of precursor powder. The size of MXene nanosheets is reported to be around tens of micrometers in maximum. The realization of large scale synthesis, especially as high-quality crystal, should be developed to explore their potentials in electronic applications. There could be exciting physics to be discovered.

The work function of MXene can be controllable by a number of different methods, and should be studied in more detail. Effects of parameters such as synthesis condition, post-synthesis heat treatment, molecular doping, and chemical treatment on work function should be thoroughly evaluated. Making MXene contact with tunable work function can enable many types of electronic devices.

Simulations suggest some MXenes such as Sc_2CT_x, Ti_2CO_2, Zr_2CO_2, and Hf_2CO_2 have large semiconductor band gaps, which opens a new field of research and potential new phenomena, such as all-MXene 2D devices, field-effect devices, Schottky junctions, and memory devices. More work should be dedicated to development of non-metallic MXenes. Recently, some simulations suggested $Mn_2C(OH)_2$, Mn_2CF_2, Ti_2NO_2, Cr_2NO_2, Mn_2NT_x MXenes as ferromagnetic half-metals. These types of MXenes are potentially important in the area of spintronics research, however experimental confirmation is required. This would also enable new functionalities and new generations of MXene devices. Double transition metal MXenes are another important class of materials that have hardly been explored. The carrier concentration and type, conductivity, work function, band gaps, and surface plasmons, can all be tuned by changing the combination of transition metals used in the compound. This can lead to drastic improvements in the performance of electronic and photonic devices based on MXenes.

Chapter 7

References

[1]   B. Anasori, M. R. Lukatskaya, Y. Gogotsi, *Nat. Rev. Mater.* **2017**, *2*, 16098.

[2]   M. Naguib, M. Kurtoglu, V. Presser, J. Lu, J. Niu, M. Heon, L. Hultman, Y. Gogotsi, M. W. Barsoum, *Adv. Mater.* **2011**, *23*, 4248.

[3]   M. Ghidiu, M. R. Lukatskaya, M.-Q. Zhao, Y. Gogotsi, M. W. Barsoum, *Nature* **2014**, *516*, 78.

[4]   G. Li, L. Tan, Y. Zhang, B. Wu, L. Li, *Langmuir* **2017**, *33*, 9000.

[5]   S. Yang, P. Zhang, F. Wang, A. G. Ricciardulli, M. R. Lohe, P. W. M. Blom, X. Feng, *Angew. Chem. Int. Ed. Engl.* **2018**, *57*, 15491.

[6]   M. Alhabeb, K. Maleski, B. Anasori, P. Lelyukh, L. Clark, S. Sin, Y. Gogotsi, *Chem. Mater.* **2017**, *29*, 7633.

[7]   M. Khazaei, M. Arai, T. Sasaki, C.-Y. Chung, N. S. Venkataramanan, M. Estili, Y. Sakka, Y. Kawazoe, *Adv. Funct. Mater.* **2013**, *23*, 2185.

[8]   U. Yorulmaz, A. Özden, N. K. Perkgöz, F. Ay, C. Sevik, *Nanotechnology* **2016**, *27*, 335702.

[9]   T. Hu, Z. Li, M. Hu, J. Wang, Q. Hu, Q. Li, X. Wang, *J. Phys. Chem. C* **2017**, *121*, 19254.

[10]   Q. Tao, M. Dahlqvist, J. Lu, S. Kota, R. Meshkian, J. Halim, J. Palisaitis, L. Hultman, M. W. Barsoum, P. O. Å. Persson, J. Rosen, *Nat. Commun.* **2017**, *8*, 14949.

[11]   R. Meshkian, M. Dahlqvist, J. Lu, B. Wickman, J. Halim, J. Thörnberg, Q. Tao, S. Li, S. Intikhab, J. Snyder, M. W. Barsoum, M. Yildizhan, J. Palisaitis, L. Hultman, P. O. Å. Persson, J. Rosen, *Adv. Mater.* **2018**, *30*, 1706409.

[12]   J. Halim, J. Palisaitis, J. Thörnberg, E. J. Moon, M. Precner, P. Eklund, P. O. Å. Persson, M. W. Barsoum, J. Rosen, *ACS Appl. Nano Mater.* **2018**, *1*, 2455.

[13]   J. Zhou, X. Zha, X. Zhou, F. Chen, G. Gao, S. Wang, C. Shen, T. Chen, C. Zhi, P. Eklund, S. Du, J. Xue, W. Shi, Z. Chai, Q. Huang, *ACS Nano* **2017**, *11*, 3841.

[14]   F. Shahzad, M. Alhabeb, C. B. Hatter, B. Anasori, S. M. Hong, C. M. Koo, Y. Gogotsi, *Science* **2016**, *353*, 1137.

[15]   K. Huang, Z. Li, J. Lin, G. Han, P. Huang, *Chem. Soc. Rev.* **2018**, *47*, 5109.

[16]   J. Ran, G. Gao, F.-T. Li, T.-Y. Ma, A. Du, S.-Z. Qiao, *Nat. Commun.* **2017**, *8*, 13907.

[17]   Z. W. Seh, K. D. Fredrickson, B. Anasori, J. Kibsgaard, A. L. Strickler, M. R. Lukatskaya, Y. Gogotsi, T. F. Jaramillo, A. Vojvodic, *ACS Energy Lett.* **2016**, *1*, 589.

[18]   H. Kim, Z. W. Wang, H. N. Alshareef, *Nano Energy* **2019**, *60*, 179.

[19]   H. Kim, B. Anasori, Y. Gogotsi, H. N. Alshareef, *Chem. Mater.* **2017**, *29*, 6472.

[20]   M. Ghidiu, J. Halim, S. Kota, D. Bish, Y. Gogotsi, M. W. Barsourm, *Chem. Mater.* **2016**, *28*, 3507.

[21]   O. Mashtalir, M. Naguib, V. N. Mochalin, Y. Dall'Agnese, M. Heon, M. W. Barsoum, Y. Gogotsi, *Nat. Commun.* **2013**, *4*, 1716.

[22]   M. R. Lukatskaya, O. Mashtalir, C. E. Ren, Y. Dall'Agnese, P. Rozier, P. L. Taberna, M. Naguib, P. Simon, M. W. Barsoum, Y. Gogotsi, *Science* **2013**, *341*, 1502.

[23]   M. W. Barsoum, T. El-Raghy, *Am. Sci.* **2001**, *89*, 334.

[24]   M. Naguib, G. W. Bentzel, J. Shah, J. Halim, E. N. Caspi, J. Lu, L. Hultman, M. W. Barsoum, *Mater. Res. Lett.* **2014**, *2*, 233.

[25]   B. Anasori, M. Dahlqvist, J. Halim, E. J. Moon, J. Lu, B. C. Hosler, E. N. Caspi, S. J. May, L. Hultman, P. Eklund, J. Rosen, M. W. Barsoum, *J. Appl. Phys.* **2015**, *118*, 094304.

[26]   B. Anasori, Y. Xie, M. Beidaghi, J. Lu, B. C. Hosler, L. Hultman, P. R. C. Kent, Y. Gogotsi, M. W. Barsourm, *ACS Nano* **2015**, *9*, 9507.

[27]   Z. Ling, C. E. Ren, M.-Q. Zhao, J. Yang, J. M. Giammarco, J. Qiu, M. W. Barsoum, Y. Gogotsi, *Proc. Natl. Acad. Sci. U.S.A.* **2014**, *111*, 16676.

[28]   M.-Q. Zhao, X. Xie, C. E. Ren, T. Makaryan, B. Anasori, G. Wang, Y. Gogotsi, *Adv. Mater.* **2017**, *29*, 1702410.

[29]   C. Zhang, B. Anasori, A. Seral-Ascaso, S.-H. Park, N. McEvoy, A. Shmeliov, G. S. Duesberg, J. N. Coleman, Y. Gogotsi, V. Nicolosi, *Adv. Mater.* **2017**, *29*, 1702678.

[30]   Z. Wang, H. Kim, H. N. Alshareef, *Adv. Mater.* **2018**, *30*, 1706656.

[31]   S. Kajiyama, L. Szabova, K. Sodeyama, H. Iinuma, R. Morita, K. Gotoh, Y. Tateyama, M. Okubo, A. Yamada, *ACS Nano* **2016**, *10*, 3334.

[32]   X. Wang, S. Kajiyama, H. Iinuma, E. Hosono, S. Oro, I. Moriguchi, M. Okubo, A. Yamada, *Nat. Commun.* **2015**, *6*, 6544.

[33]   M. Hu, Z. Li, T. Hu, S. Zhu, C. Zhang, X. Wang, *ACS Nano* **2016**, *10*, 11344.

[34]   K. Maleski, V. N. Mochalin, Y. Gogotsi, *Chem. Mater.* **2017**, *29*, 1632.

[35]   M. R. Lukatskaya, S. Kota, Z. Lin, M.-Q. Zhao, N. Shpigel, M. D. Levi, J. Halim, P.-L. Taberna, M. W. Barsoum, P. Simon, Y. Gogotsi, *Nat. Energy* **2017**, *2*, 17105.

[36]   Y. Xia, T. S. Mathis, M.-Q. Zhao, B. Anasori, A. Dang, Z. Zhou, H. Cho, Y. Gogotsi, S. Yang, *Nature* **2018**, *557*, 409.

[37]   M. A. Hope, A. C. Forse, K. J. Griffith, M. R. Lukatskaya, M. Ghidiu, Y. Gogotsi, C. P. Grey, *Phys. Chem. Chem. Phys.* **2016**, *18*, 5099.

[38]   B. Anasori, C. Shi, E. J. Moon, Y. Xie, C. A. Voigt, P. R. C. Kent, S. J. May, S. J. L. Billinge, M. W. Barsoum, Y. Gogotsi, *Nanoscale Horiz.* **2016**, *1*, 227.

[39]   J. Luo, X. Tao, J. Zhang, Y. Xia, H. Huang, L. Zhang, Y. Gan, C. Liang, W. Zhang, *ACS Nano* **2016**, *10*, 2491.

[40]   O. Mashtalir, M. R. Lukatskaya, A. I. Kolesnikov, E. Raymundo-Piñero, M. Naguib, M. W. Barsoum, Y. Gogotsi, *Nanoscale* **2016**, *8*, 9128.

[41]   X. Wang, X. Shen, Y. Gao, Z. Wang, R. Yu, L. Chen, *J. Am. Chem. Soc.* **2015**, *137*, 2715.

[42]   J. Luo, W. Zhang, H. Yuan, C. Jin, L. Zhang, H. Huang, C. Liang, Y. Xia, J. Zhang, Y. Gan, X. Tao, *ACS Nano* **2017**, *11*, 2459.

[43]   M. Ghidiu, S. Kota, J. Halim, A. W. Sherwood, N. Nedfors, J. Rosen, V. N. Mochalin, M. W. Barsoum, *Chem. Mater.* **2017**, *29*, 1099.

[44]   J. Halim, S. Kota, M. R. Lukatskaya, M. Naguib, M. Q. Zhao, E. J. Moon, J. Pitock, J. Nanda, S. J. May, Y. Gogotsi, M. W. Barsoum, *Adv. Funct. Mater.* **2016**, *26*, 3118.

[45]   M. Khazaei, A. Ranjbar, M. Arai, T. Sasaki, S. Yunoki, *J. Mater. Chem. C* **2017**, *5*, 2488.

[46]   X.-F. Yu, J.-B. Cheng, Z.-B. Liu, Q.-Z. Li, W.-Z. Li, X. Yang, B. Xiao, *RSC Adv.* **2015**, *5*, 30438.

[47]   D. Magne, V. Mauchamp, S. Célérier, P. Chartier, T. Cabioc'h, *Phys. Rev. B* **2015**, *91*, 201409.

[48]   A. N. Gandi, H. N. Alshareef, U. Schwingenschlogl, *Chem. Mater.* **2016**, *28*, 1647.

[49]   Y. Xie, P. Kent, *Phys. Rev. B* **2013**, *87*, 235441.

[50]   M. Khazaei, A. Ranjbar, M. Arai, S. Yunoki, *Phys. Rev. B* **2016**, *94*, 125152.

[51]   C. Si, K. H. Jin, J. Zhou, Z. Sun, F. Liu, *Nano Lett.* **2016**, *16*, 6584.

[52]   H. Weng, A. Ranjbar, Y. Liang, Z. Song, M. Khazaei, S. Yunoki, M. Arai, Y. Kawazoe, Z. Fang, X. Dai, *Phys. Rev. B* **2015**, *92*, 075436.

[53]   S. Kumar, U. Schwingenschlogl, *Phys. Rev. B* **2016**, *94*, 035405.

[54]   Y. Lee, Y. Hwang, Y.-C. Chung, *ACS Appl. Mater. Interfaces* **2015**, *7*, 7163.

[55]   J.-H. Liu, X. Kan, B. Amin, L.-Y. Gan, Y. Zhao, *Phys. Chem. Chem. Phys.* **2017**, *19*, 32253.

[56]   A. Chandrasekaran, A. Mishra, A. K. Singh, *Nano Lett.* **2017**, *17*, 3290.

[57]   M. Khazaei, M. Arai, T. Sasaki, M. Estili, Y. Sakka, *Phys. Chem. Chem. Phys.* **2014**, *16*, 7841.

[58]   Z. Ma, Z. Hu, X. Zhao, Q. Tang, D. Wu, Z. Zhou, L. Zhang, *J. Phys. Chem. C* **2014**, *118*, 5593.

[59]   G. Gao, A. P. O'Mullane, A. Du, *ACS Catal.* **2017**, *7*, 494.

[60]   H. Zhang, G. Yang, X. Zuo, H. Tang, Q. Yang, G. Li, *J. Mater. Chem. A* **2016**, *4*, 12913.

[61]   X.-H. Zha, Q. Huang, J. He, H. He, J. Zhai, J. S. Francisco, S. Du, *Sci. Rep.* **2016**, *6*, 27971.

[62]   J. Hu, B. Xu, C. Ouyang, S. A. Yang, Y. Yao, *J. Phys. Chem. C* **2014**, *118*, 24274.

[63]   J. Yang, X. Zhou, X. Luo, S. Zhang, L. Chen, *Appl. Phys. Lett.* **2016**, *109*, 203109.

[64]   C. Si, J. Zhou, Z. Sun, *ACS Appl. Mater. Interfaces* **2015**, *7*, 17510.

[65]   M. Khazaei, V. Wang, C. Sevik, A. Ranjbar, M. Arai, S. Yunoki, *Phys. Rev. Mater.* **2018**, *2*, 074002.

[66]   H. Lind, J. Halim, S. I. Simak, J. Rosen, *Phys. Rev. Mater.* **2017**, *1*, 044002.

[67]   X.-H. Zha, J. Zhou, K. Luo, J. Lang, Q. Huang, X. Zhou, J. S. Francisco, J. He, S. Du, *J. Phys.: Condens. Matter* **2017**, *29*, 165701.

[68]   L. Dong, H. Kumar, B. Anasori, Y. Gogotsi, V. B. Shenoy, *J. Phys. Chem. Lett.* **2017**, *8*, 422.

[69] J.-J. Zhang, L. Lin, Y. Zhang, M. Wu, B. I. Yakobson, S. Dong, *J. Am. Chem. Soc.* **2018**, *140*, 9768.

[70] M. Khazaei, M. Arai, T. Sasaki, A. Ranjbar, Y. Y. Liang, S. Yunoki, *Phys. Rev. B* **2015**, *92*, 075411.

[71] Y. Liu, H. Xiao, W. A. Goddard, *J. Am. Chem. Soc.* **2016**, *138*, 15853.

[72] H. A. Tahini, X. Tan, S. C. Smith, *Nanoscale* **2017**, *9*, 7016.

[73] M. Khazaei, A. Ranjbar, M. Ghorbani-Asl, M. Arai, T. Sasaki, Y. Liang, S. Yunoki, *Phys. Rev. B* **2016**, *93*, 205125.

[74] Y. Liu, P. Stradins, S.-H. Wei, *Sci. Adv.* **2016**, *2*, e1600069.

[75] Y. Liu, J. Guo, E. Zhu, L. Liao, S.-J. Lee, M. Ding, I. Shakir, V. Gambin, Y. Huang, X. Duan, *Nature* **2018**, *557*, 696.

[76] J. Xu, J. Shim, J. H. Park, S. Lee, *Adv. Funct. Mater.* **2016**, *26*, 5328.

[77] Z. Kang, Y. Ma, X. Tan, M. Zhu, Z. Zheng, N. Liu , L. Li, Z. Zou, X. Jiang, T. Zhai, Y. Gao, *Adv. Electron. Mater.* **2017**, *3*, 1700165.

[78] B. Z. Xu, M. S. Zhu, W. C. Zhang, X. Zhen, Z. X. Pei, Q. Xue, C. Y. Zhi, P. Shi, *Adv. Mater.* **2016**, *28*, 3333.

[79] A. D. Dillon, M. J. Ghidiu, A. L. Krick, J. Griggs, S. J. May, Y. Gogotsi, M. W. Barsoum, A. T. Fafarman, *Adv. Funct. Mater.* **2016**, *26*, 4162.

[80] K. Hantanasirisakul, M.-Q. Zhao, P. Urbankowski, J. Halim, B. Anasori, S. Kota, C. E. Ren, M. W. Barsoum, Y. Gogotsi, *Adv. Electron. Mater.* **2016**, *2*, 1600050.

[81] H. Kim, Z. Wang, M. N. Hedhili, N. Wehbe, H. N. Alshareef, *Chem. Mater.* **2017**, *29*, 2794.

[82] H. G. Kim, H. B. R. Leek, *Chem. Mater.* **2017**, *29*, 3809.

[83] J. Halim, M. R. Lukatskaya, K. M. Cook, J. Lu, C. R. Smith, L. A. Naslund, S. J. May, L. Hultman, Y. Gogotsi, P. Eklund, M. W. Barsoum, *Chem. Mater.* **2014**, *26*, 2374.

[84] V. Presser, M. Naguib, L. Chaput, A. Togo, G. Hug, M. W. Barsoum, *J. Raman Spectrosc.* **2012**, *43*, 168.

[85] T. Hu, J. M. Wang, H. Zhang, Z. J. Li, M. M. Hu, X. H. Wang, *Phys. Chem. Chem. Phys.* **2015**, *17*, 9997.

[86] P. K. Nayak, Z. Wang, H. N. Alshareef, *Adv. Mater.* **2016**, *28*, 7736.

[87]   Z. Wang, P. K. Nayak, J. A. Caraveo-Frescas, H. N. Alshareef, *Adv. Mater.* **2016**, *28*, 3831.

[88]   P. K. Nayak, Z. Wang, D. H. Anjum, M. N. Hedhili, H. N. Alshareef, *Appl. Phys. Lett.* **2015**, *106*, 103505.

[89]   J. A. Caraveo-Frescas, P. K. Nayak, H. A. Al-Jawhari, D. B. Granato, U. Schwingenschlögl, H. N. Alshareef, *ACS Nano* **2013**, *7*, 5160.

[90]   D. Hong, G. Yerubandi, H. Q. Chiang, M. C. Spiegelberg, J. F. Wager, *Crit. Rev. Solid State Mater. Sci.* **2008**, *33*, 101.

[91]   P. K. Nayak, J. A. Caraveo-Frescas, Z. Wang, M. N. Hedhili, Q. X. Wang, H. N. Alshareef, *Sci. Rep.* **2014**, *4*, 4672.

[92]   Z. Wang, H. A. Al-Jawhari, P. K. Nayak, J. A. Caraveo-Frescas, N. Wei, M. N. Hedhili, H. N. Alshareef, *Sci. Rep.* **2015**, *5*, 9617.

[93]   Z. Wang, X. He, X. X. Zhang, H. N. Alshareef, *Adv. Mater.* **2016**, *28*, 9133.

[94]   J. H. Park, F. H. Alshammari, Z. Wang, H. N. Alshareef, *Adv. Mater. Interfaces* **2016**, *3*, 1600713.

[95]   J. R. Hauser, *IEEE Trans. Educ.* **1993**, *36*, 363.

[96]   S. De, P. J. King, M. Lotya, A. O'Neill, E. M. Doherty, Y. Hernandez, G. S. Duesberg, J. N. Coleman, *Small* **2010**, *6*, 458.

[97]   S. Wang, P. K. Ang, Z. Q. Wang, A. L. L. Tang, J. T. L. Thong, K. P. Loh, *Nano Lett.* **2010**, *10*, 92.

[98]   D. Li, M. B. Muller, S. Gilje, R. B. Kaner, G. G. Wallace, *Nat Nanotechnol* **2008**, *3*, 101.

[99]   X. L. Li, G. Y. Zhang, X. D. Bai, X. M. Sun, X. R. Wang, E. Wang, H. J. Dai, *Nat Nanotechnol* **2008**, *3*, 538.

[100] G. Eda, G. Fanchini, M. Chhowalla, *Nat Nanotechnol* **2008**, *3*, 270.

[101] X. W. Wang, Z. P. Xiong, Z. Liu, T. Zhang, *Adv. Mater.* **2015**, *27*, 1370.

[102] Q. Tang, Z. Zhou, *Prog. Mater Sci.* **2013**, *58*, 1244.

[103] M. Osada, T. Sasaki, *Adv. Mater.* **2012**, *24*, 210.

[104] D. J. Norris, A. L. Efros, S. C. Erwin, *Science* **2008**, *319*, 1776.

[105] R. H. Baughman, A. A. Zakhidov, W. A. de Heer, *Science* **2002**, *297*, 787.

[106]   J. H. Choi, H. Wang, S. J. Oh, T. Paik, P. S. Jo, J. Sung, X. C. Ye, T. S. Zhao, B. T. Diroll, C. B. Murray, C. R. Kagan, *Science* **2016**, *352*, 205.

[107]   H. Kim, H. N. Alshareef, *ACS Materials Lett.* **2020**, *2*, 55.

[108]   B. Lyu, M. Kim, H. Jing, J. Kang, C. Qian, S. Lee, J. H. Cho, *ACS Nano* **2019**, *13*, 11392.

[109]   K. Montazeri, M. Currie, L. Verger, P. Dianat, M. W. Barsoum, B. Nabet, *Adv. Mater.* **2019**, *31*, 1903271.

[110]   A. Agresti, A. Pazniak, S. Pescetelli, A. Di Vito, D. Rossi, A. Pecchia, M. Auf der Maur, A. Liedl, R. Larciprete, D. V. Kuznetsov, D. Saranin, A. Di Carlo, *Nat. Mater.* **2019**, *18*, 1228.

[111]   J. Jasieniak, M. Califano, S. E. Watkins, *ACS Nano* **2011**, *5*, 5888.

[112]   F. W. Wise, *Acc. Chem. Res.* **2000**, *33*, 773.

[113]   I. Kang, F. W. Wise, *J. Opt. Soc. Am. B* **1997**, *14*, 1632.

[114]   K. J. Williams, W. A. Tisdale, K. S. Leschkies, G. Haugstad, D. J. Norris, E. S. Aydil, X. Y. Zhu, *ACS Nano* **2009**, *3*, 1532.

[115]   P. R. Brown, D. Kim, R. R. Lunt, N. Zhao, M. G. Bawendi, J. C. Grossman, V. Bulovic, *ACS Nano* **2014**, *8*, 5863.

[116]   M. J. Yuan, M. X. Liu, E. H. Sargent, *Nat. Energy* **2016**, *1*, 16016

[117]   S. Z. Bisri, C. Piliego, M. Yarema, W. G. Heiss, M. A. Loi, *Adv. Mater.* **2013**, *25*, 4309.

[118]   T. P. Osedach, N. Zhao, T. L. Andrew, P. R. Brown, D. D. Wanger, D. B. Strasfeld, L. Y. Chang, M. G. Bawendi, V. Bulovic, *ACS Nano* **2012**, *6*, 3121.

[119]   M. I. Nugraha, R. Hausermann, S. Watanabe, H. Matsui, M. Sytnyk, W. Heiss, J. Takeya, M. A. Loi, *ACS Appl. Mater. Interfaces* **2017**, *9*, 4719.

[120]   S. Z. Bisri, E. Degoli, N. Spallanzani, G. Krishnan, B. J. Kooi, C. Ghica, M. Yarema, W. Heiss, O. Pulci, S. Ossicini, M. A. Loi, *Adv. Mater.* **2014**, *26*, 5639.

[121]   A. G. Shulga, L. Piveteau, S. Z. Bisri, M. V. Kovalenko, M. A. Loi, *Adv. Electron. Mater.* **2016**, *2*, 1500467.

[122]   M. I. Nugraha, R. Hausermann, S. Z. Bisri, H. Matsui, M. Sytnyk, W. Heiss, J. Takeya, M. A. Loi, *Adv. Mater.* **2015**, *27*, 2107.

[123]   D. M. Balazs, M. I. Nugraha, S. Z. Bisri, M. Sytnyk, W. Heiss, M. A. Loi, *Appl. Phys. Lett.* **2014**, *104*, 112104.

[124]   D. Zhitomirsky, M. Furukawa, J. Tang, P. Stadler, S. Hoogland, O. Voznyy, H. Liu, E. H. Sargent, *Adv. Mater.* **2012**, *24*, 6181.

[125]   D. Balazs, D. Dirin, H. H. Fang, L. Protesescu, G. ten Brink, B. Kooi, M. Kovalenko, M. A. Loi, *ACS Nano* **2015**, *9*, 11951.

[126]   A. G. Kelly, T. Hallam, C. Backes, A. Harvey, A. S. Esmaeily, I. Godwin, J. Coelho, V. Nicolosi, J. Lauth, A. Kulkarni, S. Kinge, L. D. A. Siebbeles, G. S. Duesberg, J. N. Coleman, *Science* **2017**, *356*, 69.

[127]   M. I. Nugraha, H. Matsui, S. Watanabe, T. Kubo, R. Hausermann, S. Z. Bisri, M. Sytnyk, W. Heiss, M. A. Loi, J. Takeya, *Adv. Electron. Mater.* **2017**, *3*, 1600360.

[128]   C. F. Zhang, L. McKeon, M. P. Kremer, S. H. Park, O. Ronan, A. Seral-Ascaso, S. Barwich, C. O. Coileain, N. McEvoy, H. C. Nerl, B. Anasori, J. N. Coleman, Y. Gogotsi, V. Nicolosi, *Nat. Commun.* **2019**, *10*, 1795

[129]   Y. Z. Zhang, Y. Wang, Q. Jiang, J. K. El-Demellawi, H. Kim, H. N. Alshareef, *Adv. Mater.* **2020**, *32*, 1908486.

[130]   S. Celerier, S. Hurand, C. Garnero, S. Morisset, M. Benchakar, A. Habrioux, P. Chartier, V. Mauchamp, N. Findling, B. Lanson, E. Ferrage, *Chem. Mater.* **2019**, *31*, 454.

[131]   A. Sugahara, Y. Ando, S. Kajiyama, K. Yazawa, K. Gotoh, M. Otani, M. Okubo, A. Yamada, *Nat. Commun.* **2019**, *10*, 850.

[132]   F. A. McGuire, Y. C. Lin, K. Price, G. B. Rayner, S. Khandelwal, S. Salahuddin, A. D. Franklin, *Nano Lett.* **2017**, *17*, 4801.

[133]   J. Z. Zhang, N. Kong, S. Uzun, A. Levitt, S. Seyedin, P. A. Lynch, S. Qin, M. K. Han, W. R. Yang, J. Q. Liu, X. G. Wang, Y. Gogotsi, J. M. Razal, *Adv. Mater.* **2020**, DOI: 10.1002/adma.202001093.

[134]   S. De, T. M. Higgins, P. E. Lyons, E. M. Doherty, P. N. Nirmalraj, W. J. Blau, J. J. Boland, J. N. Coleman, *ACS Nano* **2009**, *3*, 1767.

[135]   T. M. Higgins, J. N. Coleman, *ACS Appl. Mater. Interfaces* **2015**, *7*, 16495.

[136]   V. Scardaci, R. Coull, J. N. Coleman, *Appl. Phys. Lett.* **2010**, *97*, 023114.

[137]   P. J. King, T. M. Higgins, S. De, N. Nicoloso, J. N. Coleman, *ACS Nano* **2012**, *6*, 1732.

[138]   S. Y. Lee, U. J. Kim, J. Chung, H. Nam, H. Y. Jeong, G. H. Han, H. Kim, H. M. Oh, H. Lee, H. Kim, Y.-G. Roh, J. Kim, S. W. Hwang, Y. Park, Y. H. Lee, *ACS Nano* **2016**, *10*, 6100.

[139]   Y. J. Yu, Y. Zhao, S. Ryu, L. E. Brus, K. S. Kim, P. Kim, *Nano Lett.* **2009**, *9*, 3430.

[140]   C. F. J. Zhang, S. Pinilla, N. McEyoy, C. P. Cullen, B. Anasori, E. Long, S. H. Park, A. Seral-Ascaso, A. Shmeliov, D. Krishnan, C. Morant, X. H. Liu, G. S. Duesberg, Y. Gogotsi, V. Nicolosi, *Chem. Mater.* **2017**, *29*, 4848.

[141]   M. I. Nugraha, S. Kumagai, S. Watanabe, M. Sytnyk, W. Heiss, M. A. Loi, J. Takeya, *ACS Appl. Mater. Interfaces* **2017**, *9*, 18039.

[142]   G. Yang, Y. S. Zhu, J. S. Huang, X. M. Xu, S. B. Cui, Z. W. Lu, *Opt. Express* **2019**, *27*, A1338.

[143]   M. I. Nugraha, H. Kim, B. Sun, S. Desai, F. P. G. de Arquer, E. H. Sargent, H. N. Alshareef, D. Baran, *Adv. Energy Mater.* **2019**, *9*, 1901244.

[144]   A. G. Shulga, V. Derenskyi, J. M. Salazar-Rios, D. N. Dirin, M. Fritsch, M. V. Kovalenko, U. Scherf, M. A. Loi, *Adv. Mater.* **2017**, *29*, 1701764.

[145]   J. H. Cho, J. Lee, Y. He, B. Kim, T. P. Lodge, C. D. Frisbie, *Adv. Mater.* **2008**, *20*, 686.

[146]   T. M. Higgins, S. Finn, M. Matthiesen, S. Grieger, K. Synnatschke, M. Brohmann, M. Rother, C. Backes, J. Zaumseil, *Adv. Funct. Mater.* **2019**, *29*, 1804387.

[147]   Z. F. Lin, D. Barbara, P. L. Taberna, K. L. Van Aken, B. Anasori, Y. Gogotsi, P. Simon, *J. Power Sources* **2016**, *326*, 575.

[148]   B. P. Bloom, M. N. Mendis, E. Wierzbinski, D. H. Waldeck, *J. Mater. Chem. C* **2016**, *4*, 704.

[149]   C. E. Ren, K. B. Hatzell, M. Alhabeb, Z. Ling, K. A. Mahmoud, Y. Gogotsi, *J. Phys. Chem. Lett.* **2015**, *6*, 4026.

[150]   S. Hong, F. W. Ming, Y. Shi, R. Y. Li, I. S. Kim, C. Y. Y. Tang, H. N. Alshareef, P. Wang, *ACS Nano* **2019**, *13*, 8917.

[151]   M. A. Hines, G. D. Scholes, *Adv. Mater.* **2003**, *15*, 1844.

[152]   A. Sarycheva, A. Polemi, Y. L. Liu, K. Dandekar, B. Anasori, Y. Gogotsi, *Sci. Adv.* **2018**, *4*, eaau0920.

[153]   M. Naguib, J. Halim, J. Lu, K. M. Cook, L. Hultman, Y. Gogotsi, M. W. Barsoum, *J. Am. Chem. Soc.* **2013**, *135*, 15966.

[154]   O. Mashtalir, M. R. Lukatskaya, M. Q. Zhao, M. W. Barsoum, Y. Gogotsi, *Adv. Mater.* **2015**, *27*, 3501.

[155]   M. Q. Zhao, C. E. Ren, Z. Ling, M. R. Lukatskaya, C. F. Zhang, K. L. Van Aken, M. W. Barsoum, Y. Gogotsi, *Adv. Mater.* **2015**, *27*, 339.

[156]   Q. Tang, Z. Zhou, P. W. Shen, *J. Am. Chem. Soc.* **2012**, *134*, 16909.

[157]   H. Liu, C. Y. Duan, C. H. Yang, W. Q. Shen, F. Wang, Z. F. Zhu, *Sens. Actuators, B* **2015**, *218*, 60.

[158]   F. Wang, C. H. Yang, C. Y. Duan, D. Xiao, Y. Tang, J. F. Zhu, *J. Electrochem. Soc.* **2015**, *162*, B16.

[159]   Q. R. Zhang, J. Teng, G. D. Zou, Q. M. Peng, Q. Du, T. F. Jiao, J. Y. Xiang, *Nanoscale* **2016**, *8*, 7085.

[160]   K. Rasool, M. Helal, A. Ali, C. E. Ren, Y. Gogotsi, K. A. Mahmoud, *ACS Nano* **2016**, *10*, 3674.

[161]   J. C. Lei, X. Zhang, Z. Zhou, *Frontiers of Physics* **2015**, *10*, 276.

[162]   Y. L. Bai, X. D. He, R. G. Wang, *J. Raman Spectrosc.* **2015**, *46*, 784.

[163]   Y. Xu, Z. X. Gan, S. C. Zhang, *Phys. Rev. Lett.* **2014**, *112*, 226801.

[164]   Y. L. Chen, J. G. Analytis, J. H. Chu, Z. K. Liu, S. K. Mo, X. L. Qi, H. J. Zhang, D. H. Lu, X. Dai, Z. Fang, S. C. Zhang, I. R. Fisher, Z. Hussain, Z. X. Shen, *Science* **2009**, *325*, 178.

[165]   C. Hu, C. C. Lai, Q. Tao, J. Lu, J. Halim, L. Sun, J. Zhang, J. Yang, B. Anasori, J. Wang, Y. Sakka, L. Hultman, P. Eklund, J. Rosen, M. W. Barsoum, *Chem. Commun.* **2015**, *51*, 6560.

[166]   B. Anasori, J. Halim, J. Lu, C. A. Voigt, L. Hultman, M. W. Barsoum, *Scripta Mater.* **2015**, *101*, 5.

[167]   M. Naguib, R. R. Unocic, B. L. Armstrong, J. Nanda, *Dalton Trans.* **2015**, *44*, 9353.

[168]   K. D. Fredrickson, B. Anasori, Z. W. Seh, Y. Gogotsi, A. Vojvodic, *J. Phys. Chem. C* **2016**, *120*, 28432.

[169]   W. Sirisaksoontorn, A. A. Adenuga, V. T. Remcho, M. M. Lerner, *J. Am. Chem. Soc.* **2011**, *133*, 12436.

[170]   C. F. Zhang, M. Beidaghi, M. Naguib, M. R. Lukatskaya, M. Q. Zhao, B. Dyatkin, K. M. Cook, S. J. Kim, B. Eng, X. Xiao, D. H. Long, W. M. Qiao, B. Dunn, Y. Gogotsi, *Chem. Mater.* **2016**, *28*, 3937.

[171]   N. J. Lane, M. Naguib, V. Presser, G. Hug, L. Hultman, M. W. Barsoum, *J. Raman Spectrosc.* **2012**, *43*, 954.

[172]   M. Naguib, O. Mashtalir, M. R. Lukatskaya, B. Dyatkin, C. F. Zhang, V. Presser, Y. Gogotsi, M. W. Barsoum, *Chem. Commun.* **2014**, *50*, 7420.

[173]   D. Yang, A. Velamakanni, G. Bozoklu, S. Park, M. Stoller, R. D. Piner, S. Stankovich, I. Jung, D. A. Field, C. A. Ventrice, R. S. Ruoff, *Carbon* **2009**, *47*, 145.

[174]   O. Chaix-Pluchery, A. Thore, S. Kota, J. Halim, C. Hu, J. Rosen, T. Ouisse, M. W. Barsoum, *J. Raman Spectrosc.* **2017**, *48*, 631.

[175]   L. H. Li, *Comput. Mater. Sci.* **2016**, *124*, 8.

[176]   C. Si, J. You, W. Shi, J. Zhou, Z. Sun, *J. Mater. Chem. C* **2016**, *4*, 11524.

[177]   M. Naguib, T. Saito, S. Lai, M. S. Rager, T. Aytug, M. P. Paranthaman, M. Q. Zhao, Y. Gogotsi, *RSC Adv.* **2016**, *6*, 72069.

[178]   Y. Xie, M. Naguib, V. N. Mochalin, M. W. Barsoum, Y. Gogotsi, X. Q. Yu, K. W. Nam, X. Q. Yang, A. I. Kolesnikov, P. R. C. Kent, *J. Am. Chem. Soc.* **2014**, *136*, 6385.

[179]   Y. F. Li, Y. C. Ding, B. Xiao, Y. H. Cheng, *Phys. Lett. A* **2016**, *380*, 3748.

[180]   J. Y. Oh, J. H. Lee, S. W. Han, S. S. Chae, E. J. Bae, Y. H. Kang, W. J. Choi, S. Y. Cho, J. O. Lee, H. K. Baik, T. Il Lee, *Energy Environ. Sci.* **2016**, *9*, 1696.

[181]   K. Kuroki, R. Arita, *J. Phys. Soc. Jpn.* **2007**, *76*, 083707.

[182]   K. Mori, H. Sakakibara, H. Usui, K. Kuroki, *Phys. Rev. B* **2013**, *88*, 075141.

[183]   H. Usui, K. Suzuki, K. Kuroki, S. Nakano, K. Kudo, M. Nohara, *Phys. Rev. B* **2013**, *88*, 075140.

[184]   H. Lu, P. G. Burke, A. C. Gossard, G. Zeng, A. T. Ramu, J. H. Bahk, J. E. Bowers, *Adv. Mater.* **2011**, *23*, 2377.

[185]   Y. Chen, M. He, B. Liu, G. C. Bazan, J. Zhou, Z. Liang, *Adv. Mater.* **2017**, *29*, 1604752.

[186]   H. Kim, Z. W. Wang, M. N. Hedhili, N. Wehbe, H. N. Alshareef, *Chem. Mater.* **2017**, *29*, 2794.

APPENDICES

Accomplishments

1. Peer-reviewed Journal Papers:

1st Author & co-1st Author Papers

[1] "Quantum Dot Electrical Double Layer Transistors with MXene Electrical Contacts"
Hyunho Kim, M. I. Nugraha, X. Guan, Z. Wang, M. K. Hota, X. Xu, T. Wu, D. Baran, T. D. Anthopoulos, H. N. Alshareef
Under preparation

[2] "MXetronics: MXene-Enabled Electronic and Photonic Devices"
Hyunho Kim, H. N. Alshareef
ACS Materials Lett. **2020**, 2 (1), 55-70

[3] "MXetronics: Electronic and photonic applications of MXenes"
Hyunho Kim, Z. Wang, H. N. Alshareef
Nano Energy **2019**, 60, 179-197

[4] "Oxide Thin-Film Electronics using All-MXene Electrical Contacts"
Z. Wang, **Hyunho Kim** (co-1st author), H. N. Alshareef
Adv. Mater. **2018**, 30 (15), 1706656

[5] "Thermoelectric Properties of Two-Dimensional Molybdenum-Based MXenes"
Hyunho Kim, B. Anasori, Y. Gogotsi, H. N. Alshareef
Chem. Mater. **2017**, 29 (15), 6472-6479

[6] "Oxidant-dependent thermoelectric properties of undoped ZnO films by atomic layer deposition"
Hyunho Kim, Z. Wang, M. N. Hedhili, N. Wehbe, H. N. Alshareef
Chem. Mater. **2017**, 29 (7), 2794-2802

Co-Author Papers

[7] "Substitutional Doping Enables Low Thermal Conductivity N-type Quantum Dot-in-Matrix Films"
M. I. Nugraha, B. Sun, **H. Kim**, A. El-Labban, S. Desai, N. Chaturvedi, Y. Hou, F. P. Garcia de Arquer, H. N. Alshareef, E. H. Sargent, D. Baran
Under preparation

[8] "Iontronics Using V_2CT_x MXene-Derived Metal-Organic Framework Solid Electrolytes"
X. Xu, H. Wu, X. He, M. K. Hota, Z. Liu, S. Zhuo, **Hyunho Kim**, X. Zhang, H. N. Alshareef
ACS Nano **2020**, accepted (DOI: 10.1021/acsnano.0c02497)

[9] "Photothermoelectric Response of $Ti_3C_2T_x$ MXene Confined Ion Channels"
S. Hong, G. Zou, **Hyunho Kim**, D. Huang, P. Wang, H. N. Alshareef
ACS Nano **2020**, 14 (7), 9042-9049

[10] "MXene Printing and Patterned Coating for Device Applications"
Y.-Z. Zhang, Y. Wang, Q. Jiang, J. K. El-Demellawi, **Hyunho Kim**, H. N. Alshareef
Adv. Mater. **2020**, 32 (21), 1908486

[11] "Ultrasound-Driven Two-Dimensional $Ti_3C_2T_x$ MXene Hydrogel Generator"
K. H. Lee, Y.-Z. Zhang, Q. Jiang, **Hyunho Kim**, A. A. Alkenawi, H. N. Alshareef
ACS Nano **2020**, 14 (3), 3199-3207

[12] "Titanium Carbide MXene Nucleation Layer for Epitaxial Growth of High-Quality GaN Nanowires on Amorphous Substrates"
A. Prabaswara, **Hyunho Kim**, J.-W. Min, R. C. Subedi, D. H. Anjum, B. Davaasuren, K. Moore, M. Conroy, S. Mitra, I. S. Roqan, T. K. Ng, H. N. Alshareef, B. S. Ooi
ACS Nano **2020**, 14 (2), 2202-2211

[13] "Heteroatom-Mediated Interactions between Ruthenium Single Atoms and an MXene Support for Efficient Hydrogen Evolution"
V. Ramalingam, P. Varadhan, H. C. Fu, **Hyunho Kim**, D. L. Zhang, S. M. Chen, L. Song, D. Ma, Y. Wang, H. N. Alshareef, J. H. He
Adv. Mater. **2019**, 31 (48), 1903841

[14] "Highly Passivated n-Type Colloidal Quantum Dots for Solution-Processed Thermoelectric Generators with Large Output Voltage"
M. I. Nugraha, **Hyunho Kim**, B. Sun, S. Desai, F. P. G. de Arquer, E. H. Sargent, H.

N. Alshareef, D. Baran
Adv. Energy Mater. **2019**, 9 (28), 1901244

[15] "MXene-Contacted Silicon Solar Cells with 11.5% Efficiency"
H. C. Fu, V. Ramalingam, **Hyunho Kim**, C. H. Lin, X. S. Fang, H. N. Alshareef, J. H. He
Adv. Energy Mater. **2019**, 9 (22), 1900180

[16] "Highly Stretchable and Air-Stable PEDOT: PSS/Ionic Liquid Composites for Efficient Organic Thermoelectrics"
S. Kee, **Hyunho Kim**, S. H. K. Paleti, A. El Labban, M. Neophytou, A. H. Emwas, H. N. Alshareef, D. Baran
Chem. Mater. **2019**, 31 (9), 3519-3526

[17] "Low-Temperature-Processed Colloidal Quantum Dots as Building Blocks for Thermoelectrics"
M. I. Nugraha, **Hyunho Kim**, B. Sun, M. A. Haque, F. P. G. de Arquer, D. R. Villalva, A. El-Labban, E. H. Sargent, H. N. Alshareef, D. Baran
Adv. Energy Mater. **2019**, 9 (13), 1803049

[18] "MXenes stretch hydrogel sensor performance to new limits"
Y.-Z. Zhang, K. H. Lee, D. H. Anjum, R. Sougrat, Q. Jiang, **Hyunho Kim**, H. N. Alshareef
Sci. Adv. **2018**, 4 (6), eaat0098